Annals of Mathematics Studies

Number 3

THE CONSISTENCY OF THE AXIOM OF CHOICE AND OF THE GENERALIZED CONTINUUM-HYPOTHESIS WITH THE AXIOMS OF SET THEORY

BY

Kurt Gödel

PRINCETON, NEW JERSEY
PRINCETON UNIVERSITY PRESS
1940

ISBN 0-691-07927-7

Eighth Printing 1970

In the second and subsequent printings a few misprints and minor inaccuracies have been corrected. Note in particular the references to the highly important developments in the foundations of set theory that have occurred since 1960.

PRINTED IN THE UNITED STATES OF AMERICA

Notes by George W. Brown of lectures delivered at the Institute for Advanced Study, Princeton, New Jersey, during the autumn term, 1938-1939.

CONTENTS

INTRODUCTION

In these lectures it will be proved that the axiom of choice and Cantor's generalised continuum-hypothesis (i.e. the proposition that $2^{\aleph_\alpha} = \aleph_{\alpha+1}$ for any α) are consistent with the other axioms of set theory if these axioms are consistent. The system Σ of axioms for set theory which we adopt includes the axiom of substitution [cf. A. Fraenkel, Zehn Vorlesungen über die Grundlegung d. Mengenlehre Wiss. u. Hyp. 31 (Teubner, 1927), p. 115] and the axiom of "Fundierung" [cf. E. Zermelo, Fund. Math. 16, p. 31] but of course does not include the axiom of choice. It is essentially due to P. Bernays [cf. Journ. Symb. Log. 2, p. 65] and is equivalent with v. Neumann's system S* + VI [cf. J. reine angew. Math. 160, p. 227], if the axiom of choice is left out, or, to be more exact, if axiom III3* is replaced by axiom III3. What we shall prove is that if a contradiction from the axiom of choice and the generalised continuum-hypothesis were derived in Σ, it could be transformed into a contradiction obtained from the axioms of Σ alone. This result is obtained by constructing within Σ (i.e. using only the primitive terms and axioms of Σ) a model Δ for set theory with the following properties: 1) the propositions which say that the axioms of Σ hold for Δ are theorems demonstrable in Σ, 2) the propositions which say that the axiom of choice and the generalised continuum-hypothesis hold in Δ are likewise demonstrable in Σ. In fact there is a much stronger proposition* which can be proved to hold in Δ and which has other interesting consequences besides the axiom of choice and the generalised continuum-hypothesis (cf. p. 47).

In order to define Δ and to prove the above properties of it from the axioms of Σ, it is necessary first to develop abstract set theory to a certain extent from the axioms of Σ. This is done in Chapters II-IV. Although the definitions and theorems are mostly stated in logistic symbols, the theory developed is not to be considered as a formal system but as an axiomatic theory in which the meaning and the properties of the logical symbols are presupposed to be known. However to everyone familiar with mathematical logic it will be clear that the proofs could be formalised, using only the rules of Hilbert's "Engerer Funktionenkalkul". In several places (in particular for the "general existence theorem" on p. 8 and the notions of "relativisation" and of "absoluteness" on p. 42) we are concerned with metamathematical considerations about the notions and propositions of the system Σ. However the only purpose of these general metamathematical considerations is to show how the proofs for theorems of a certain kind can be accomplished by

---

* See Note 1 of Notes added in the second printing, p. 67.

a general method. And since applications to only a finite number of instances are necessary for proving the properties 1) and 2) of the model Δ, the general metamathematical considerations could be left out entirely, if one took the trouble to carry out the proofs separately for any instance.[1]

In the first introductory part about set theory in general (i.e. in Chapters II-IV) not all proofs are carried out in detail, since many of them can be literally transferred from non-axiomatic set theory and, moreover, an axiomatic treatment on a very similar basis has been given by J. v. Neumann [Math. Z. 27, p. 669].

For the logical notions we use the following symbols: (X), ($\exists$X), $\sim$, $\cdot$, $\vee$, $\supset$, $\equiv$, $=$, (E!X), which mean respectively: for all X, there is an X, not, and, or, implies, equivalence, identity, there is exactly one X. X=Y means that X and Y are the same object. "For all X" is also expressed by free variables in definitions and theorems.

The system Σ has in addition to the ε-relation two primitive notions, namely "class " and "set". Classes are what appear in Zermelo's formulation [Math. Ann. 65, p. 263] as "definite Eigenschaften". However, in the system Σ (unlike Zermelo's) it is stated explicitly by a special group of axioms (group B on p. 5), how definite Eigenschaften are to be constructed. Classes represent at the same time relations between sets, namely a class A represents the relation which subsists between x and y if the ordered pair $\langle xy \rangle$ (defined in 1.12) is an element of A. The same ε-relation is used between sets and sets, and sets and classes. The axiom of extensionality (Fraenkel's Bestimmtheitsaxiom) is assumed for both sets and classes and a class for which there exists a set having the same elements is identified with this set, so that every set is a class.[2] On the other hand a class B which is not a set (e.g. the universal class) can never occur as an element owing to axiom A.2, i.e., BεX is then always false (but meaningful).

---

[1] In particular also the complete inductions used in the proofs of theorems 1.16, M1, M2 are needed only up to a certain definite integer, say 20.

[2] Similarly, v. Neumann in Math. Z. 27.

Note. Dots are also used to replace brackets in the well-known manner.

CHAPTER I

THE AXIOMS OF ABSTRACT SET THEORY

Our primitive notions are: class, denoted by $\mathfrak{Cls}$; set, denoted by $\mathfrak{M}$; and the diadic relation ε between class and class, class and set, set and class, or set and set. The primitive notions appear in context as follows:

$\mathfrak{Cls}(A)$, A is a class
$\mathfrak{M}(A)$, A is a set
$X\varepsilon Y$, $X\varepsilon y$, $x\varepsilon Y$, $x\varepsilon y$,

where the convention is made that X, Y, Z, ... , are variables whose range consists of all the classes, and that x, y, z, ... , are variables whose range is all sets.

The axioms fall into four groups, A, B, C, D.

Group A.

1. $\mathfrak{Cls}(x)$
2. $X\varepsilon Y.\supset.\mathfrak{M}(X)$
3. $(u)[u\varepsilon X.\equiv.u\varepsilon Y].\supset.X=Y$
4. $(x)(y)(\exists z)[u\varepsilon z.\equiv:u=x.\vee.u=y]$

Axiom 1 in the group above states that every set is a class. A class which is not a set is called a proper class, i.e.

1. Dfn $\mathfrak{Pr}(X)\equiv\sim\mathfrak{M}(X)$.

Axiom 2 says that every class which is a member of some class is a set. Axiom 3 is the principle of extensionality, that is, two classes are the same, if their elements are the same. Axiom 4 provides for the existence of the set whose members are just x and y , for any sets x and y . Moreover, this set is defined uniquely for given x and y , by axiom 3. The element z defined by 4 is called the non-ordered pair of x and y , denoted by $\{xy\}$, i.e.

1.1 Dfn $u\varepsilon\{xy\}\equiv(u=x\vee u=y)$.
1.11 Dfn $\{x\}=\{xx\}$

$\{x\}$ is the set whose sole member is x .

1.12 Dfn $\langle xy\rangle=\{\{x\}\{xy\}\}$

$\langle xy\rangle$ is called the <u>ordered pair</u> of x and y. We have the following theorem:

1.13 $\langle xy\rangle=\langle uv\rangle . \supset : x=u . y=v$,

that is, two ordered pairs are equal if and only if the corresponding elements of each are equal. In this sense, $\langle xy\rangle$ is an ordered pair. The proof of this theorem is not difficult. [cf. P. Bernays, Journ. Symb. Log. 2, p. 69]

The <u>ordered triple</u> may now be defined in terms of the ordered pair.

1.14 Dfn $\langle xyz\rangle=\langle x\langle yz\rangle\rangle$

The corresponding theorem holds for the ordered triple. The <u>n-tuple</u> can be defined by induction as follows:

1.15 Dfn $\langle x_1x_2\ldots x_n\rangle=\langle x_1\langle x_2\ldots x_n\rangle\rangle$.

This gives the theorem

1.16 $\langle x_1\ldots x_n\langle x_{n+1}\ldots x_{n+p}\rangle\rangle=\langle x_1\ldots x_nx_{n+1}\ldots x_{n+p}\rangle$,

which is proved by induction on n.
In order that $\langle\ \rangle$ be defined for any number of arguments it is convenient to put

1.17 Dfn $\langle x\rangle=x$,

which entails the equation 1.16 also for the case $p=1$.
We also define <u>inclusion</u> $\subseteq$ and <u>proper inclusion</u> $\subset$.

1.2 Dfn $X\subseteq Y . \equiv . (u)[u\varepsilon X . \supset . u\varepsilon Y]$; $X\subset Y . \equiv : X\subseteq Y . X\neq Y$.

A class is said to be <u>empty</u> if it has no members; "X is empty" is denoted by "$\mathfrak{Em}(X)$", i.e.

1.22 Dfn $\mathfrak{Em}(X) \equiv (u)\sim u\varepsilon X$.

If X and Y have no members in common, we write "$\mathfrak{Ex}(X,Y)$", that is, "X and Y are <u>mutually exclusive</u>", i.e.

1.23 Dfn $\mathfrak{Ex}(X,Y) \equiv (u)\sim(u\varepsilon X . u\varepsilon Y)$.

X is said to be <u>one-many</u> (<u>single-valued</u>), denoted by "$\mathfrak{Un}(X)$", if for any u there exists at most one v such that $\langle vu\rangle\varepsilon X$, that is:

1.3 Dfn $\mathfrak{Un}(X) \equiv (u,v,w)[\langle vu\rangle \varepsilon X . \langle wu\rangle \varepsilon X : \supset . v=w]$.

The axioms of the second group are concerned with the existence of classes:

Group B.

1. $(\exists A)(x,y)[\langle xy\rangle \varepsilon A . \equiv . x\varepsilon y]$
2. $(A)(B)(\exists C)(u)[u\varepsilon C . \equiv : u\varepsilon A . u\varepsilon B]$
3. $(A)(\exists B)(u)[u\varepsilon B . \equiv . \sim (u\varepsilon A)]$
4. $(A)(\exists B)(x)[x\varepsilon B . \equiv . (\exists y)(\langle yx\rangle \varepsilon A)]$
5. $(A)(\exists B)(x,y)[\langle yx\rangle \varepsilon B . \equiv . x\varepsilon A]$
6. $(A)(\exists B)(x,y)[\langle xy\rangle \varepsilon B . \equiv . \langle yx\rangle \varepsilon A]$
7. $(A)(\exists B)(x,y,z)[\langle xyz\rangle \varepsilon B . \equiv . \langle yzx\rangle \varepsilon A]$
8. $(A)(\exists B)(x,y,z)[\langle xyz\rangle \varepsilon B . \equiv . \langle xzy\rangle \varepsilon A]$

Axiom B1 is called axiom of the ε-relation, B2 axiom of intersection, B3 axiom of the complement, B4 axiom of the domain, B5 axiom of the direct product (because it provides essentially for the existence of $V \times A$, V being the universal class), B6-8 axioms of inversion.[3] Note that the class A in axiom B1 and the class B in axiom B5-8 are not uniquely determined, since nothing is said about those sets which are not pairs (triples), whether or not they belong to A (B). On the other hand in axioms B2-4 the classes C and B are uniquely determined (owing to axiom A.3). These uniquely determined classes in B2-4 are denoted respectively by A.B, -A, $\mathfrak{D}(A)$ and called *intersection* of A, B, *complement* of A, *domain* of A, respectively. Thus A.B, -A, $\mathfrak{D}(A)$ are defined by the following properties.

1.4 Dfn $x\varepsilon A.B \equiv x\varepsilon A . x\varepsilon B$
1.41 Dfn $x\varepsilon -A \equiv \sim x\varepsilon A$
1.5 Dfn $x\varepsilon \mathfrak{D}(A) \equiv (\exists y)\langle yx\rangle \varepsilon A$

The third group of axioms is concerned with the existence of sets.

Group C.

1. $(\exists a)\{\sim \mathfrak{Em}(a) . (x)[x\varepsilon a . \supset . (\exists y)[y\varepsilon a . x \subset y]]\}$
2. $(x)(\exists y)(u,v)[u\varepsilon v . v\varepsilon x : \supset . u\varepsilon y]$
3. $(x)(\exists y)[u \subseteq x . \supset . u\varepsilon y]$
4. $(x,A)\{\mathfrak{Un}(A) . \supset . (\exists y)(u)[u\varepsilon y . \equiv . (\exists v)[v\varepsilon x . \langle uv\rangle \varepsilon A]]\}$

Axiom 1 is the so-called *axiom of infinity*. There is a non-void set a such that given any element x of a , there is another element y of a , of which x is a proper subset. According to axiom 2, for any set x there is a set y including the sum of all elements of x . Axiom 3 provides for the

[3] Note that axioms B7 and B8 have as consequences similar theorems for any permutation of a triple.

existence of a set including the set of all subsets of x. Axiom 4 is the <u>axiom of substitution</u>★; for any set x and any single-valued A, there is a set y whose elements are just those sets which bear the relation defined by A to members of x [Instead of C4, Zermelo used the <u>Aussonderungsaxiom</u>:

$$(x,A)(\exists y)(u)[u \varepsilon y . \equiv : u \varepsilon x . u \varepsilon A] \quad ,$$

that is, there is a set whose members are just those elements of x which have the property A].

The following axiom [proved consistent by v. Neumann, J. reine angew. Math. 160, p. 227] is not indispensable, but it simplifies considerably the later work.

Axiom D: $\sim \mathfrak{Em}(A) . \supset . (\exists u)[u \varepsilon A . \mathfrak{Ex}(u,A)]$,

that is, any non-void class A has some element with no members in common with A.[4] It is a consequence of D that:

$$1.6 \quad \sim x \varepsilon x$$

for if there were such an x, x would be a common element of x and $\{x\}$, but, by D, taking $\{x\}$ for A, x can have no element in common with $\{x\}$. Likewise:

$$1.7 \quad \sim [x \varepsilon y . y \varepsilon x] \quad .$$

This follows by considering $\{xy\}$ in an analogous way.

The following axiom is the axiom of choice.✢

Axiom E. $(\exists A)\{\mathfrak{Un}(A) . (x)[\sim \mathfrak{Em}(x) . \supset . (\exists y)[y \varepsilon x . \langle yx \rangle \varepsilon A]]\}$

This is a very strong form of the <u>axiom of choice</u>, since it provides for the simultaneous choice, by a single relation, of an element from each set of the universe under consideration. From this form of the axiom, one can prove that the whole universe of sets can be well-ordered. This stronger form of the axiom, if consistent with the other axioms, implies, of course, that a weaker form is also consistent.

---

[4] This axiom is equivalent to the non-existence of infinite descending sequences of sets◇(i.e. such that $x_{i+1} \varepsilon x_i$) where however the term "sequence" refers only to sequences representable by sets of the system under consideration. That is (using the definitions 4.65, 7.4, 8.41 below) axiom D is (owing to the axioms of the groups A, B, C, E) equivalent to the proposition $\sim(\exists y)(i)[y`(i \dot{+} 1) \varepsilon y`i]$.

★ See Note 2
◇ See Note 3
✢ See Note 4
} of Notes added to the second printing, p. 67.

The system of axioms of groups A, B, C, D is called Σ.[5] If a theorem is stated without further specification it means that it follows from Σ. If the axiom E is needed for a theorem or a definition its number is marked by a *.

---

[5] The most important differences between Σ and the system of P. Bernays [Journ. Symb. Log. 2, 65] are:

1. Bernays does not identify sets and classes having the same extension.

2. Bernays assumes a further axiom requiring the existence of the class of all $\{x\}$, which allows B7 and B8 to be replaced by one axiom.

Axiom D is essentially due to v. Neumann [cf. J. reine angew. Math. 160, p. 231, Axiom VI 4], whose formulation however is more complicated, because his system has other primitive terms. The concise formulation used in the text is due to P. Bernays.

CHAPTER II

EXISTENCE OF CLASSES AND SETS

We now define the metamathematical notion of a primitive propositional function (abbreviated ppf). A ppf will be a meaningful formula containing only variables, symbols for special classes $A_1, \dots, A_k$, ε, and logical operators, and such that all bound variables are set variables. For example,

$$(u)(u\varepsilon X. \supset . u\varepsilon A) \text{ and } (u)[u\varepsilon x. \equiv .(v)[v\varepsilon u. \supset . v\varepsilon y]]$$

are ppf. A formula is non-primitive if (X) or $(\exists X)$ occurs.

More precisely, ppf can be defined recursively as follows: Let $\Pi, \Gamma, \dots,$ denote variables or special classes, then

(1) $\Pi\varepsilon\Gamma$ is a ppf.

(2) If φ and ψ are ppf, then so are $\sim\varphi$ and $\varphi.\psi$.

(3) If φ is a ppf, then $(\exists x)\varphi$ is a ppf, and any result of replacing x by another set variable is a ppf.

(4) Only formulas obtained by 1, 2, 3 are ppf.

Logical operators different from $\sim$, $\cdot$, $\exists$, need not be mentioned since they can be defined in terms of these three.

The following metatheorem says that the extension of any ppf is represented by a class:

M1. General Existence Theorem: If $\varphi(x_1, \dots, x_n)$ is a ppf containing no free variables other than $x_1, \dots, x_n$ (not necessarily all these) then there exists a class A, such that for any sets $x_1, \dots, x_n$,

$$\langle x_1 \dots x_n \rangle \varepsilon A. \equiv . \varphi(x_1, \dots, x_n) \quad .$$

For the proof of this theorem, several preliminary results are needed.

By means of the axioms on intersection and complement, it is possible to prove the existence of a universal class V and a null class 0. Because of the axiom of extensionality, 0 and V are uniquely determined by the properties

2.1 Dfn $(x)\sim(x\varepsilon 0)$,

2.2 Dfn $(x)x\varepsilon V$.

As a consequence of axiom B5, the axiom of the direct product, and B6, the axiom of the inverse relation, we have

2.3 $(A)(\exists B)(x,y)[\langle xy\rangle\varepsilon B. \equiv .x\varepsilon A]$.

The following three theorems are also consequences of B5, B7, and B8.

2.31 $(A)(\exists B)(x,y,z)[\langle zxy\rangle\varepsilon B. \equiv .\langle xy\rangle\varepsilon A]$
2.32 $(A)(\exists B)(x,y,z)[\langle xzy\rangle\varepsilon B. \equiv .\langle xy\rangle\varepsilon A]$
2.33 $(A)(\exists B)(x,y,z)[\langle xyz\rangle\varepsilon B. \equiv .\langle xy\rangle\varepsilon A]$

For example, the first of these theorems is proved by substituting an ordered pair for the second member in the ordered pair appearing in B5, rewriting the variables properly. The other two are obtained by applying to 2.31 the axioms of inversion (B7 and B8).

Substituting $\langle x_1 x_2 \ldots x_n\rangle$ for x in B5 in a similar way, we get

2.4 $(A)(\exists B)(y,x_1,\ldots,x_n)[\langle yx_1 x_2\ldots x_n\rangle\varepsilon B. \equiv .\langle x_1\ldots x_n\rangle\varepsilon A]$.

From this, by iteration,

2.41 $(A)(\exists B)(y_1,\ldots,y_k,x_1,\ldots,x_n)[\langle y_1\ldots y_k x_1\ldots x_n\rangle\varepsilon B.$
$$\equiv .\langle x_1\ldots x_n\rangle\varepsilon A]$$.

Similarly,

2.5 $(A)(\exists B)(y_1,\ldots,y_k,x_1,\ldots,x_n)[\langle x_1 y_1\ldots y_k x_2\ldots x_n\rangle\varepsilon B.$
$$\equiv .\langle x_1\ldots x_n\rangle\varepsilon A]$$.

This may be obtained by iteration from the case $k=1$, and this case in turn is a special case of 2.32 obtained by substituting $\langle x_2\ldots x_n\rangle$ for y and applying theorem 1.16.

The following theorems are derived in an analogous fashion, by substituting $\langle y_1\ldots y_k\rangle$ for z and y respectively in 2.33, 2.3, and applying 1.16,

2.6 $(A)(\exists B)(x_1,x_2,y_1,\ldots,y_k)[\langle x_1 x_2 y_1\ldots y_k\rangle\varepsilon B. \equiv .\langle x_1 x_2\rangle\varepsilon A]$
2.7 $(A)(\exists B)(x,y_1,\ldots,y_k)[\langle xy_1\ldots y_k\rangle\varepsilon B. \equiv .x\varepsilon A]$

The next (and for the present, the last) theorem is a generalization of axiom B4, the axiom of the domain, and is obtained by substituting, in B4, $\langle x_2\ldots x_n\rangle$ for x .

2.8 $(A)(\exists B)(x_2,\ldots,x_n)[\langle x_2\ldots x_n\rangle\varepsilon B. \equiv .(\exists x_1)[\langle x_1\ldots x_n\rangle\varepsilon A]]$

In particular $B=\mathfrak{D}(A)$ satisfies this equivalence.

In the proof of the general existence theorem, it can be assumed that none of the special classes A_i appears as the first argument of the ε-relation, because $A_i\varepsilon\Gamma$ can be replaced by

$(\exists x)(x=A_i . x\varepsilon\Gamma)$ (by axiom A2) and $x=A_i$ can be replaced by $(u)[u\varepsilon x \equiv u\varepsilon A_i]$ (by axiom A3).

The proof of M1 is an inductive one, the induction taking place on the number of logical operators in φ.

Case 1. φ has no logical operators.
In this case φ has one of two possible forms, $x_r \varepsilon x_s$ and $x_r \varepsilon A_k$, where $1 \leq r, s \leq n$. If φ is of the form $x_r \varepsilon x_s$, we must show that there exists a class A such that $\langle x_1 \ldots x_n \rangle \varepsilon A . \equiv . x_r \varepsilon x_s$. If $r=s$, take as A the null class 0, since, by 1.6, $\sim(x_r \varepsilon x_r)$. If $r \neq s$, φ must be either of the form $x_p \varepsilon x_q$ or $x_q \varepsilon x_p$, where $p<q$. For $x_p \varepsilon x_q$, axiom B1 provides for the existence of an F such that $\langle x_p x_q \rangle \varepsilon F . \equiv . x_p \varepsilon x_q$. For $x_q \varepsilon x_p$, B1 followed by B6 provides for the existence of an F such that

$$\langle x_p x_q \rangle \varepsilon F . \equiv . x_q \varepsilon x_p \quad .$$

Therefore, in either case there is an F such that

$$\langle x_p x_q \rangle \varepsilon F . \equiv . \varphi(x_1, \ldots, x_n) \quad .$$

Now, by 2.6, there is an F_1 with the property:

$$\langle x_p x_q x_{q+1} \ldots x_n \rangle \varepsilon F_1 . \equiv . \langle x_p x_q \rangle \varepsilon F \quad .$$

Then by 2.5 there exists F_2 such that

$$\langle x_p \ldots x_n \rangle \varepsilon F_2 . \equiv . \langle x_p x_q x_{q+1} \ldots x_n \rangle \varepsilon F_1 \quad ,$$

and finally, by 2.41 there exists a class A such that

$$\langle x_1 \ldots x_n \rangle \varepsilon A . \equiv . \langle x_p \ldots x_n \rangle \varepsilon F_2 \quad .$$

Combining these equivalences the result is:

$$\langle x_1 \ldots x_n \rangle \varepsilon A . \equiv . \varphi(x_1, \ldots, x_n) \quad .$$

Now suppose φ is of the form $x_r \varepsilon A_k$. By 2.3, there is an F such that $\langle x_r x_{r+1} \rangle \varepsilon F . \equiv . \varphi(x_1, \ldots, x_n)$. (If $r=n$, use axiom B5 to get $\langle x_{r-1} x_r \rangle \varepsilon F . \equiv . \varphi(x_1, \ldots, x_n)$.) Now, as above, by means of theorems 2.6 and 2.41, combining the resulting equivalences establishes the existence of A.

Case 2. φ has m logical operators $(m>0)$.
Then φ has one of the following three forms:

$$\text{(a)} \sim \psi \; ; \quad \text{(b)} \; \psi \cdot \chi \; ; \quad \text{(c)} \; (\exists x)\vartheta \quad .$$

The hypothesis of the induction is that for all ppfs $\psi(x_1, \ldots, x_n)$ with $m_1<m$ logical operators, and such that no A_i appears in the context $A_i \varepsilon \Gamma$, there exists an A with the properties required by the theorem. ψ, χ, and ϑ are ppfs with fewer than m logical operators. ψ and χ have no other free variables than at most $x_1, \ldots, x_n$, whereas ϑ has no other free variables than at most $x, x_1, \ldots, x_n$, and A_i cannot appear in the context $A_i \varepsilon \Gamma$ in ψ, χ or ϑ, because it does not appear in φ in this context. Therefore, by the hypothesis of the induction, there exist classes B, C, D, such that

$$\langle x_1 \ldots x_n \rangle \varepsilon B . \equiv . \psi(x_1, \ldots, x_n) \quad ,$$
$$\langle x_1 \ldots x_n \rangle \varepsilon C . \equiv . \chi(x_1, \ldots, x_n) \quad ,$$
$$\langle x x_1 \ldots x_n \rangle \varepsilon D . \equiv . \vartheta(x, x_1, \ldots, x_n) \quad .$$

For (a) take A as $-B$, since, by axiom B3,

$$\langle x_1 \ldots x_n \rangle \varepsilon -B . \equiv . \sim(\langle x_1 \ldots x_n \rangle \varepsilon B) \quad ,$$

so that $\langle x_1 \ldots x_n\rangle \varepsilon -B. \equiv . \sim\psi(x_1,\ldots,x_n)$,
that is $\langle x_1 \ldots x_n\rangle \varepsilon -B. \equiv .\varphi(x_1,\ldots,x_n)$.
For (b) take A as B·C, since by axiom B2,

$$\langle x_1 \ldots x_n\rangle \varepsilon B\cdot C. \equiv :\langle x_1 \ldots x_n\rangle \varepsilon B. \langle x_1 \ldots x_n\rangle \varepsilon C \quad ,$$

that is $\langle x_1 \ldots x_n\rangle \varepsilon B\cdot C. \equiv :\psi(x_1,\ldots,x_n) . \chi(x_1,\ldots,x_n)$
therefore $\langle x_1 \ldots x_n\rangle \varepsilon B\cdot C. \equiv .\varphi(x_1,\ldots,x_n)$.
For (c), take A as the domain $\mathfrak{D}(D)$, since, by theorem 2.8

$$\langle x_1 \ldots x_n\rangle \varepsilon \mathfrak{D}(D). \equiv .(\exists x)[\langle x x_1 \ldots x_n\rangle \varepsilon D]$$

therefore $\langle x_1 \ldots x_n\rangle \varepsilon \mathfrak{D}(D). \equiv .(\exists x)\vartheta(\langle x x_1,\ldots,x_n)$,
so that $\langle x_1 \ldots x_n\rangle \varepsilon \mathfrak{D}(D). \equiv .\varphi(x_1,\ldots,x_n)$.
This completes the proof of the general existence theorem for primitive propositional functions.

The general existence theorem is a <u>metatheorem</u>, that is, a theorem about the system, not in the system, and merely indicates once and for all, how the formal derivation would proceed in the system for any given ppf.

So far, the existence theorem is proved only for ppfs, but the use of symbols introduced by definition yields a wider class of propositional functions for which it would be desirable to have the existence theorem valid. With this in view, examine the defined symbols introduced thus far. They may be classified into four types, as follows:

1. <u>Particular classes</u>: 0, V, ... ,
2. <u>Notions</u>: $\mathfrak{M}(X)$, $\mathfrak{Pr}(X)$, $\mathfrak{Un}(X)$, $X \subseteq Y$, ... ,
3. <u>Operations</u>: $-X$, $\mathfrak{D}(X)$, $X\cdot Y$, ... ,
4. <u>Kinds of variables</u>: x, X, ... , (defined by notions).

Henceforth it is to be required that all operations and notions be meaningful, that is, defined, for all classes as arguments. This has been the case hitherto except for the pairs $\{xy\}$ and $\langle xy\rangle$, and the n-tuples, which were defined for sets only. The extension for classes as arguments can be accomplished simply by replacing the free set-variables by class-variables in the definitions, i.e.,

3.1 Dfn $(u)[u\varepsilon\{XY\}. \equiv :u=X.\vee .u=Y]$,
3.11 Dfn $\{X\}=\{XX\}$,
3.12 Dfn $\langle XY\rangle=\{\{X\}\{XY\}\}$, etc.

By these definitions e.g. $\{XY\}$ is either $\{XY\}$ or $\{X\}$ or $\{Y\}$ or 0 according to whether both or one or none of X, Y are sets. The same procedure of extension is to be applied in definitions 4.211, 4.65, 6.31, 7.4, where the notions (or operations) under consideration are originally defined only if certain arguments are sets.[6]

See Notes 5 and 6 of Notes added to the second printing, p. 68.
[6] Note that in all these definitions it is absolutely unimpor-

The following metamathematical ideas will be useful. A <u>term</u> is defined inductively so that (1) any variable is a term, and any symbol denoting a special class is a term; (2) if $\mathfrak{A}$ is an operation with n arguments and $\Gamma_1,\dots,\Gamma_n$ are terms, then $\mathfrak{A}(\Gamma_1,\dots,\Gamma_n)$ is a term; (3) there are no terms other than those obtainable from (1) and (2). If $\mathfrak{B}$ is a notion with n arguments and $\Gamma_1,\dots,\Gamma_n$ are terms, then $\mathfrak{B}(\Gamma_1,\dots,\Gamma_n)$ is said to be a <u>minimal propositional function</u> or <u>minimal formula</u>. A <u>propositional function</u> may be defined recursively as any result of combining minimal propositional functions by means of the logical operators: $\sim$, $\vee$, $\cdot$, $\supset$, $\equiv$ and quantifiers for any kind of variables.

For each of the four types of symbols there is a corresponding kind of definition.

1. A special class A is introduced by a <u>defining postulate</u> $\varphi(A)$, where φ is a propositional function containing only previously defined symbols and it has to be proved first that there is exactly one class A, such that $\varphi(A)$.

2. A notion $\mathfrak{B}$ is introduced by the stipulation

$$\mathfrak{B}(X_1,\dots,X_n) \equiv \varphi(X_1,\dots,X_n) \quad ,$$

where φ is a propositional function containing only previously defined symbols.

3. An operation $\mathfrak{A}$ is introduced by a <u>defining postulate</u>

$$(X_1,\dots,X_n)\varphi(\mathfrak{A}(X_1,\dots,X_n),X_1,\dots,X_n) \quad ,$$

where φ is a propositional function containing only previously defined symbols and it has first to be proved that

$$(X_1,\dots,X_n)(E!Y)\varphi(Y,X_1,\dots,X_n) \quad .$$

4. A variable $\mathfrak{x}$ is introduced by a stipulation that for any propositional function φ, $(\mathfrak{x})\varphi(\mathfrak{x})$ means:

$$(X)[\mathfrak{B}(X) \supset \varphi(X)]$$

and $(\exists\mathfrak{x})\varphi(\mathfrak{x})$ means:

$$(\exists X)[\mathfrak{B}(X)\cdot\varphi(X)] \quad ,$$

where $\mathfrak{B}$ is a previously defined notion the extension of which is called the <u>range of the variable</u> $\mathfrak{x}$.

Special classes, notions and operations are sometimes referred to by the common name "concepts".◇

All definitions so far introduced are of this type: $\mathfrak{B}$ is called a <u>normal notion</u> if there is a ppf φ such that

$$\mathfrak{B}(X_1,\dots,X_n) .\equiv. \varphi(X_1,\dots,X_n) \quad ,$$

$\mathfrak{A}$ is called a <u>normal operation</u> if there is a ppf φ such that

$$Y\varepsilon\mathfrak{A}(X_1,\dots,X_n) .\equiv. \varphi(Y,X_1,\dots,X_n) \quad ,$$

and a <u>variable is called normal</u> if its range consists of the

---

tant how the notions or operations under consideration are defined for proper classes as arguments.* The only purpose of defining them at all for this case is to simplify the metamathematical concepts of "term" and "propositional function" defined on p.12 and the formulation of Theorems M2-M6.

◇ See Note 7 }
* See Note 7 } of Notes added to the second printing, p. 68.

elements of a class. The propositional function $\varphi(X_1,\ldots,X_n)$ is called normal if it contains only normal operations, normal notions, and normal bound variables, and a term is called normal if it contains only normal operations.

M2. Any normal propositional function is equivalent to some ppf and therefore M1 holds also for any normal propositional function $\varphi(X_1,\ldots,X_n)$.

Proof: Let $\varphi(X_i,\ldots,X_n)$ be the given normal propositional function. Since φ contains only normal bound variables all bound variables not set variables can be replaced by set variables, e.g., $(\exists \mathfrak{x})\chi(\mathfrak{x})$ by $(\exists x)[x\varepsilon A.\chi(x)]$, where A defines the range of the variable $\mathfrak{x}$. Next, for any notion $\mathfrak{A}$ occurring in φ, since it is normal, the minimal propositional function $\mathfrak{A}(\Gamma_1,\ldots,\Gamma_n)$ can be replaced by the equivalent $\psi(\Gamma_1,\ldots,\Gamma_n)$, where $\psi(X_1,\ldots,X_n)$ is a ppf. Then the only remaining notion is the ε-relation. Again all contexts of the form $\Gamma\varepsilon\Delta$ where Γ is not a set variable can be removed by the method explained on p.10 after theorem 2.8, leaving only minimal formulas of the form $u\varepsilon\Gamma$. But Γ, if not a variable or a special class, is of the form $\mathfrak{B}(\Gamma_1,\ldots,\Gamma_n)$, where $\mathfrak{B}$ is a normal operation. But $u\varepsilon\mathfrak{B}(\Gamma_1,\ldots,\Gamma_n)$ can be replaced by $\psi(u,\Gamma_1,\ldots,\Gamma_n)$, where the ppf ψ is such that $u\varepsilon\mathfrak{B}(\Gamma_1,\ldots,\Gamma_n).\equiv.\psi(u,\Gamma_1,\ldots,\Gamma_n)$. In this way, φ is reduced, getting all operations out. The final result of such reductions can be nothing other than a ppf.

This completes the proof that M1 is valid for normal propositional functions. It remains only to verify that all concepts introduced so far are normal. This will be done by constructing for each of the corresponding expressions $Y\varepsilon\mathfrak{A}(X_1,\ldots,X_n)$ and $\mathfrak{B}(X_1,\ldots,X_n)$ equivalent propositional functions containing only notions, operations and bound variables previously shown to be normal. These propositional functions are then equivalent to ppfs by theorem M2.

$X\varepsilon Y$; ε is normal, since $X\varepsilon Y$ is itself a ppf.
$X=Y.\equiv.(u)[u\varepsilon X.\equiv.u\varepsilon Y]$
$\mathfrak{M}(X).\equiv.(\exists u)(u=X)$
$\mathfrak{Pr}(X).\equiv.\sim\mathfrak{M}(X)$
$Z\,\varepsilon\{XY\}.\equiv:(Z=X.\vee.Z=Y).\mathfrak{M}(Z)$
$Z\,\varepsilon\langle XY\rangle.\equiv.Z\,\varepsilon\{\{X\}\{XY\}\}$ and similarly for triples, etc.
$X\subseteq Y.\equiv.(u)(u\varepsilon X.\supset.u\varepsilon Y)$
$X\subset Y.\equiv.(u)(u\varepsilon X.\supset.u\varepsilon Y).\sim(X=Y)$
$\mathfrak{Un}\,(X).\equiv.(u,v,w)[\langle uv\rangle\varepsilon X.\langle wv\rangle\varepsilon X:\supset.u=w]$
$X\varepsilon(-A).\equiv:\mathfrak{M}(X).\sim(X\varepsilon A)$
$X\varepsilon A\cdot B.\equiv:X\varepsilon A.X\varepsilon B$
$X\,\varepsilon\mathfrak{D}(A).\equiv:\mathfrak{M}(X).(\exists y)[\langle yX\rangle\varepsilon A]$
$\mathfrak{Em}\,(X).\equiv.\sim(\exists u)(u\varepsilon X)$
$\mathfrak{Ex}(XY).\equiv.\sim(\exists u)(u\varepsilon X.u\varepsilon Y)$

The general existence theorems M1, M2 (and likewise the later theorems M3-M6) are frequently used in these lectures without being quoted explicitly.

The particular classes $A_1,\ldots,A_k$ that may appear in the normal propositional function $\varphi(x_1,\ldots,x_n)$ are entirely arbitrary, and may therefore be replaced by the general class variables $X_1,\ldots,X_k$, so that the existence theorem takes the form:

M3. $$(X_1,\ldots,X_k)(\exists A)(x_1,\ldots,x_n)[\langle x_1\ldots x_n\rangle\varepsilon A. \equiv.\varphi(x_1,\ldots,x_n,X_1,\ldots,X_k)],$$

if φ is normal.

The definitions that follow are mostly based on the existence theorem in this form. In each application of M3 it is apparent upon inspection that φ is normal.

The direct (outer) product $A\times B$ is defined by the postulate:

4.1 Dfn $(x)[x\varepsilon A\times B. \equiv .(\exists y,z)[x=\langle yz\rangle :y\varepsilon A.z\varepsilon B]]$

A and B are considered as the constant classes in this application of M3, which assures the existence of $A\times B$ for all A and B. That $A\times B$ is unique is guaranteed by the axiom of extensionality.

4.11 Dfn $A^2=A\times A$

4.12 Dfn $A^3=A\times(A^2)$

A^4, A^5, ... , are defined similarly. Thus V^2 is the class of all ordered pairs, V^3 is the class of all ordered triples, etc. Since every triple is a pair, it follows that

4.13 $V^3\subseteq V^2$

Relations are to be defined as classes of ordered pairs, triadic relations as classes of ordered triples, etc.

4.2 Dfn $\mathfrak{Rel}(X). \equiv .X\subseteq V^2$

4.21 Dfn $\mathfrak{Rel}_3(X). \equiv .X\subseteq V^3$

and similarly for all $n\geq 2$. "$\mathfrak{Rel}(X)$" may be written as "$\mathfrak{Rel}_2(X)$".

If A is a relation, then $\langle xy\rangle\varepsilon A$ is read "x bears the relation A to y", and may be written xAy, i.e.

4.211 Dfn $xAy. \equiv .\langle xy\rangle\varepsilon A$.

Relations can be thought of as many-valued functions, so that xAy may be read also as "x is a value of A for the argument y" or "x is an image of y by A", or "y is an original of x, with respect to A". As a corollary of the axiom of extensionality, there is a principle of extensionality for relations:

4.22 $\mathfrak{Rel}(X).\mathfrak{Rel}(Y):\supset:(u,v)[\langle uv\rangle\varepsilon X. \equiv .\langle uv\rangle\varepsilon Y].\supset .X=Y$.

The extensionality principle for relations holds also for n-adic relations, in a similar manner. As a result, the existence theorem takes the form:

M4. Given a normal propositional function $(x_1,\ldots,x_n)$, there is exactly one n-adic relation A such that

$$(x_1,\ldots,x_n)[\langle x_1\ldots\ldots x_n\rangle \varepsilon A. \equiv .\varphi(x_1,\ldots,x_n)] \quad .$$

The proof is immediate. Take an arbitrary class A' satisfying the condition, and take A as $A'\ V^n$. A is an n-adic relation and is unique because of the principle of extensionality, 4.22.

A, as defined by M4, is denoted by $\hat{x}_1,\ldots,\hat{x}_n[\varphi(x_1,\ldots,x_n)]$. If $\alpha_1,\ldots,\alpha_n$ are normal variables, $\hat{\alpha}_1,\ldots,\hat{\alpha}_n[\varphi(\alpha_1,\ldots,\alpha_n)]$ is by definition the same as $\hat{x}_1,\ldots,\hat{x}_n[\varphi(x_1,\ldots,x_n).x_1\varepsilon C,\ldots,x_n\varepsilon C]$, where C is the range of the variables α_i. [Note that the symbol ^ belongs to none of the four kinds of symbols introduced on p. 11, therefore it must not be used in definitions or in applications of M2-M6.]

The ε-relation E and the identity relation I may be defined by means of M4.

4.3 Dfn $\mathfrak{Rel}(E).(u,v)[\langle uv\rangle\varepsilon E. \equiv .u\varepsilon v]$

4.31 Dfn $\mathfrak{Rel}(I).(u,v)[\langle uv\rangle\varepsilon I. \equiv .u=v]$

I is the class of all pairs $\langle uu\rangle$.

The following definitions 4.4, 4.41, 4.411 of the <u>converse relations</u> correspond to the axioms B6,7,8.

4.4 Dfn $\mathfrak{Rel}[\mathfrak{Cnv}(X)].(u,v)[\langle uv\rangle\ \varepsilon\ \mathfrak{Cnv}(X). \equiv .\langle vu\rangle\varepsilon X]$

4.41 Dfn $\mathfrak{Rel}_3[\mathfrak{Cnv}_2(X)].(u,v,w)[\langle uvw\rangle\ \varepsilon\ \mathfrak{Cnv}_2(X). \equiv .\langle vwu\rangle\varepsilon X]$

4.411 Dfn $\mathfrak{Rel}_3[\mathfrak{Cnv}_3(X)].(u,v,w)[\langle uvw\rangle\ \varepsilon\ \mathfrak{Cnv}_3(X). \equiv .\langle uwv\rangle\varepsilon X]$

4.412 Dfn $\mathfrak{Cnv}(X)$ is also denoted by $\mathfrak{Cnv}_1(X)$, X^{-1}, and $\check{X}$.

The binary Boolean operations "+" and "-" are defined in terms of "·" and the complement "-":

4.42 Dfn $X+Y=-[(-X)\cdot(-Y)]$,

4.43 Dfn $X-Y=X\cdot(-Y)$,

4.44 Dfn $\mathfrak{W}(X)=\mathfrak{D}(X^{-1})$.

$\mathfrak{W}(X)$ is called <u>domain of values</u> of X.

The relation "<u>A confined to B</u>" is written "$A\upharpoonright B$".

4.5 Dfn $A\upharpoonright B=A\cdot(V\times B)$

$A\upharpoonright B$ consists of all elements of A which are ordered pairs with second member from B. In that sense, "$A\upharpoonright B$" is "A confined to B", since the arguments of A are restricted to lie in B. This gives the theorem:

4.51 $\mathfrak{D}(A \upharpoonright B) = B \cdot \mathfrak{D}(A)$.
4.512 Dfn $B \upharpoonleft A = A \cdot (B \times V)$
4.52 Dfn $B“X = \mathfrak{W}(B \upharpoonright X)$

B"X is the class of all images by B of elements of X.

4.53 Dfn $\langle xy \rangle \varepsilon R|S . \equiv . (\exists z)(xRz . zSy)] . \mathfrak{Rel}(R|S)$
4.6 Dfn $\mathfrak{Un}_2(X) . \equiv : \mathfrak{Un}(X) . \mathfrak{Un}(X^{-1})$

$\mathfrak{Un}_2(X)$ means X is one to one, that is, the relation $X \cdot V^2$ is one to one. If X is a relation, and is single-valued, X is said to be a function.

4.61 Dfn $\mathfrak{Fnc}(X) . \equiv : \mathfrak{Rel}(X) . \mathfrak{Un}(X)$.

A function X whose domain is A is called a function over A.

4.63 Dfn $X \mathfrak{Fn} A . \equiv : \mathfrak{Fnc}(X) . \mathfrak{D}(X) = A$

$A'x$ (the A of x) denotes the y such that $\langle yx \rangle \varepsilon A$, if

that y exists and is unique; if y does not exist or is not unique, $A'x=0$. Hence the defining postulate for $A'x$ reads as follows:

4.65 Dfn $(E!y)[\langle yx \rangle \varepsilon A] . \supset . \langle A'x, x \rangle \varepsilon A :$
$\sim(E!y)[\langle yx \rangle \varepsilon A] . \supset . A'x = 0 .: \mathfrak{M}(A'x)$.

The extensionality principle for relations (4.22) gives the following extensionality principle for functions:

4.67 $X \mathfrak{Fn} A . Y \mathfrak{Fn} A : \supset : (u)[u \varepsilon A . \supset : X'u = Y'u] . \supset . X = Y$.

M5. If $\psi(u_1, \ldots, u_n)$ is a normal term, if $B \subseteq V^n$ and if $\langle u_1 \ldots u_n \rangle \varepsilon B . \supset . \mathfrak{M}(\psi(u_1, \ldots, u_n))$
then there exists exactly one function C over B such that
$C'\langle u_1, \ldots, u_n \rangle = \psi(u_1, \ldots, u_n)$ for $\langle u_1 \ldots u_n \rangle \varepsilon B$.
Proof: Define C by the condition:
$\langle u u_1 \ldots u_n \rangle \varepsilon C . \equiv : u = \psi(u_1, \ldots, u_n) . \langle u_1, \ldots, u_n \rangle \varepsilon B$.
Since the right hand side is normal there is an (n+1)-adic relation C satisfying the condition, by M4. C obviously satisfies the conditions of the theorem.

M5 may be generalized as follows:

M6. If $B_1, \ldots, B_k$ are mutually exclusive, $B_i \subseteq V^n$, and if $\psi_1, \ldots, \psi_k$ are normal terms such that $\mathfrak{M}(\psi_i(u_1, \ldots, u_n))$ for $\langle u_1, \ldots, u_n \rangle \varepsilon B_i$ then there exists exactly one function C over $B_1 + B_2 + \ldots + B_n$ such that $C'\langle u_1 \ldots u_n \rangle = \psi_i(u_1, \ldots, u_n)$ for

$\langle u_1 \ldots u_n \rangle \varepsilon B_i$, $i=1,2,\ldots,k$.

We now define five special functions $P_1,\ldots,P_5$ by the following postulates:

4.71 Dfn $P_1{}^{\text{c}}\langle xy\rangle = x . P_1\ \mathfrak{Fn}\ V^2$,
4.72 Dfn $P_2{}^{\text{c}}\langle xy\rangle = y . P_2\ \mathfrak{Fn}\ V^2$,
4.73 Dfn $P_3{}^{\text{c}}\langle xy\rangle = \langle yx\rangle . P_3\ \mathfrak{Fn}\ V^2$,
4.74 Dfn $P_4{}^{\text{c}}\langle xyz\rangle = \langle zxy\rangle . P_4\ \mathfrak{Fn}\ V^3$,
4.75 Dfn $P_5{}^{\text{c}}\langle xyz\rangle = \langle xzy\rangle . P_5\ \mathfrak{Fn}\ V^3$.

Existence and unicity of $P_1,\ldots,P_5$ follow from M5.

4.8 Dfn $u\varepsilon\mathfrak{S}(X) . \equiv . (\exists v)[u\varepsilon v . v\varepsilon X]$

$\mathfrak{S}(X)$ is called the sum of X. The following results are immediate:

4.81 $\mathfrak{S}\{xy\} = x+y$,
4.82 $\mathfrak{S}\{x\} = x$,
4.83 $\mathfrak{S}(X) = E``X$.

Now define $\mathfrak{P}(X)$, the power class of X, the class of subsets of X.

4.84 Dfn $u\varepsilon\mathfrak{P}(X) . \equiv . u \subseteq X$

Some of the operations defined have monotonicity properties, e.g.,

4.85 $X \subseteq Y . \supset : \mathfrak{D}(X) \subseteq \mathfrak{D}(Y)$.

It is easily verified that $\mathfrak{W}$, $\mathfrak{S}$, $\mathfrak{P}$, and $\mathfrak{Cnv}_1$ have similar properties. Also

4.86 $A \subseteq B . X \subseteq Y : \supset . A``X \subseteq B``Y$.

1, $\upharpoonright$, $+$, $\cdot$, and $\times$ have similar properties.

We also have some distributivities, such as

4.87 $(A\times B)\cdot(C\times D) = (A\cdot C)\times(B\cdot D)$.

This leads to the special case

4.871 $(A\times V)\cdot(V\times B) = A\times B$.

Likewise

4.88 $\mathfrak{S}(X+Y) = \mathfrak{S}(X)+\mathfrak{S}(Y)$,
4.89 $\mathfrak{S}(X\cdot Y) \subseteq \mathfrak{S}(X)\cdot\mathfrak{S}(Y)$.

The following theorems result from definitions 4.71-4.75, and are immediate upon inspection.

4.91 $\mathfrak{W}(A)=P_1$“ A
4.92 $\mathfrak{D}(A)=P_2$“ A
4.93 $\mathfrak{Cnv}(A)=P_3$“ A
4.94 $\mathfrak{Cnv}_2(A)=P_4$“ A
4.95 $\mathfrak{Cnv}_3(A)=P_5$“ A
4.96 $V\times A=\breve{P}_2$“ A

The proof for the normality of the notions and operations introduced above and also of those introduced later is contained on p. 62 .

The results obtained thus far depended on the first two groups of axioms. Theorems on the existence of sets depend, however, on the later axioms. The following theorem depends on axiom C4, the axiom of substitution.

5.1 $\mathfrak{Un}(A) . \mathfrak{M}(X) : \supset . \mathfrak{M}(A$“ $X)$

Proof: Since $\mathfrak{M}(X)$, there is a set y , by C4, whose elements are just those sets which bear the relation $A \cdot V^2$ to members of X, that is, $(u)[u\varepsilon y . \equiv . u\varepsilon A$“$X]$, so that, by the axiom of extensionality, y is identical with A“X. Therefore $\mathfrak{M}(A$“$X)$.

5.11 $\mathfrak{M}(X) . \supset . \mathfrak{M}(X \cdot Y)$

Proof: Substitute $I \upharpoonright Y$ for A in 5.1, obtaining $\mathfrak{M}[(I \upharpoonright Y)$“$X]$. But $(I \upharpoonright Y)$“$X=X \cdot Y$.

5.12 $\mathfrak{M}(X) \cdot Y \subseteq X : \supset . \mathfrak{M}(Y)$

Proof: $Y \subseteq X . \supset . Y=X \cdot Y$. Now, by 5.11, the theorem is proved.

5.121 $\mathfrak{M}(X) . \supset . \mathfrak{M}(\mathfrak{P}(X))$

Proof: Axiom C3 provides for the existence of a y such that $\mathfrak{P}(X) \subseteq y$. Therefore by 5.12, $\mathfrak{M}(\mathfrak{P}(X))$.

5.122 $\mathfrak{M}(X) . \supset . \mathfrak{M}(\mathfrak{S}(X))$

Proof: This is proved similarly by using axiom C2 and 5.12.

5.13 $\mathfrak{M}(X) . \mathfrak{M}(Y) : \supset . \mathfrak{M}(X+Y)$

Proof: If X, Y are sets, we have $X+Y=\mathfrak{S}\{XY\}$ and by axiom A4, $\{XY\}$ is a set. Therefore by 5.122, $\mathfrak{M}(X+Y)$. The next three theorems are proved by 5.1 using 4.91-4.95.

5.14 $\mathfrak{M}[\mathfrak{D}(x)]$
5.15 $\mathfrak{M}[\mathfrak{Cnv}_i(x)]$ $(i=1,2,3)$
5.16 $\mathfrak{M}[\mathfrak{W}(x)]$

From 5.14 and M5 it follows that there is a function Do such that:

5.17 Dfn $Do\text{'}x=\mathfrak{D}(x) . Do\,\mathfrak{Fn}\,V$.
5.18 $\mathfrak{M}(x \times y)$

Proof: The members of $x \times y$ are the pairs $\langle uv \rangle$, where $u \varepsilon x$, $v \varepsilon y$. In particular, then, u and v are elements of $x+y$, so that $\{u\}$ and $\{uv\}$ are subsets of $x+y$. Therefore $\{\{u\}\{uv\}\}$ is a subset of $\mathfrak{P}(x+y)$, that is, $\langle uv \rangle \subseteq \mathfrak{P}(x+y)$, so that $\langle uv \rangle \varepsilon \mathfrak{P}[\mathfrak{P}(x+y)]$, i.e., $x \times y \subseteq \mathfrak{P}[\mathfrak{P}(x+y)]$. Therefore $\mathfrak{M}(x \times y)$, by 5.121 and 5.12 and 5.13.

5.19 $F\,\mathfrak{Fn}\,x . \supset . \mathfrak{M}(F)$

Proof: $F\,\mathfrak{Fn}\,x . \supset . F \subseteq (F``x) \times x$, therefore $\mathfrak{M}(F)$, by 5.1, 5.18, 5.12

5.2 $\mathfrak{Un}(F) . \supset . \mathfrak{M}(F \upharpoonright x)$

Proof: $F \upharpoonright x$ is a function over $\mathfrak{D}(F \upharpoonright x)$ and $\mathfrak{D}(F \upharpoonright x) \subseteq x$ hence a set. Hence the theorem holds by 5.19.

5.3 $\mathfrak{M}(0)$

Proof: $0 \subseteq x$, therefore $\mathfrak{M}(0)$, by 5.12

5.31 $\sim \mathfrak{M}(V)$

Proof: $x \varepsilon V$, therefore if $\mathfrak{M}(V)$ we would have $V \varepsilon V$, but this is impossible, by 1.6.

5.4 $\mathfrak{Pr}(X) . \supset . \mathfrak{Pr}(\mathfrak{S}(X))$

Proof: Suppose $\mathfrak{M}(\mathfrak{S}(X))$, then $\mathfrak{M}(\mathfrak{P}(\mathfrak{S}(X)))$, but $X \subseteq \mathfrak{P}(\mathfrak{S}(X))$, therefore $\mathfrak{M}(X)$, contrary to the hypothesis.

Similarly:

5.41 $\mathfrak{Pr}(X) . \supset . \mathfrak{Pr}(\mathfrak{P}(X))$,
5.42 $\mathfrak{Pr}(X) . \supset . \mathfrak{Pr}(X+Y)$,
5.43 $\mathfrak{Pr}(X) . \sim \mathfrak{Em}(Y) : \supset . \mathfrak{Pr}(X \times Y)$.
Proof: $X \subseteq \mathfrak{S}[\mathfrak{S}(X \times Y)]$, if $Y \neq 0$.

5.44 $\mathfrak{Un}_2(F) . X \subseteq \mathfrak{D}(F) : \supset : \mathfrak{Pr}(X) . \supset . \mathfrak{Pr}(F``X)$ , that is, a one to one image of a proper class is a proper class. The proof follows from the fact that $X \subseteq \breve{F}``(F``X)$ , if $X \subseteq \mathfrak{D}(F)$ . Therefore if $F``X$ were a set X would also be a set by 5.1 and 5.12.

5.45 $\mathfrak{Pr}(A) . \supset . \mathfrak{Pr}(A-x)$

This follows from the inclusion $A \subseteq (A-x)+x$, and 5.13.

CHAPTER III

ORDINAL NUMBERS

Ordinal numbers may now be defined, with the aid of some preliminary definitions.

6.1 Dfn $Y \mathfrak{Con} X . \equiv . X^2 \subseteq Y + Y^{-1} + I$,

that is, Y is connex in X, if for any pair of distinct elements u, v of X, either $\langle uv \rangle \varepsilon Y$ or $\langle vu \rangle \varepsilon Y$.

6.11 Dfn Y is called transitive in X if, for all elements u, v, w of X,

$\langle uv \rangle \varepsilon Y . \langle vw \rangle \varepsilon Y : \supset \langle uw \rangle \varepsilon Y$.

6.12 Dfn Y is called asymmetric in X if, for no elements u, v of X,

$\langle uv \rangle \varepsilon Y . \langle vu \rangle \varepsilon Y$.

6.2 Dfn $X \mathfrak{We} Y . \equiv : Y \mathfrak{Con} X . (U)[U \neq 0 . U \subseteq X : \supset . (\exists v)[v \varepsilon U . U \cdot Y^{\text{``}}\{v\} = 0]]$,

that is, X is well ordered by Y if Y is connex in X and any nonvoid subset U of X has a first element with respect to the ordering Y, since $U \cdot Y^{\text{``}}\{v\} = 0$ says that there is no member of U which bears Y to v . Note that the symbol $X \mathfrak{We} Y$ here introduced is not normal, because of the bound variable U.

6.21 If $X \mathfrak{We} Y$, then Y is transitive and asymmetric in X.

Proof: Y is asymmetric in X, since if xYy and yYx the class {xy} has no first element. In order to prove the transitivity in X, suppose xYy and yYz , then $x \neq z$ because of the asymmetry, hence either xYz or zYx . Consider $U = \{x\} + \{y\} + \{z\}$. If zYx , U will have no first element, therefore xYz .

6.3 Dfn $X \mathfrak{Sect}_R Y . \equiv : X \subseteq Y . [Y \cdot (R^{\text{``}}X) \subseteq X]$

that is, X is an R-section of Y if all R-predecessors in Y of members of X also belong to X.

6.30 Dfn X is called a proper R-section of Y, if it is an R-section of Y and $\neq$Y.

See Note 8 of Notes added to the second printing, p. 68.

6.31 Dfn $\mathfrak{Seg}_R(X,u)=X\cdot R``\{u\}$,

that is, if $u\varepsilon X$ *the R-segment of X generated by u* is the class of elements of X which are R-predecessors of u .

6.32 $\mathfrak{Seg}_R(X,u)$ is an R-section of X if $u\varepsilon X$ and if R is transitive in X.

Therefore

6.33 If $X\,\mathfrak{We}\,R$, then any R-segment generated by an element of X is an R-section.

Conversely, if $X\,\mathfrak{We}\,R$ and Y is a proper R-section of X, then Y is an R-segment of X, namely the one generated by the first element of X-Y.

If R is a one to one relation with domain A and converse domain B, then R is called an *isomorphism from A to B with respect to S and T* if for any pair u, v of A such that uSv the corresponding elements of B are in the relation T, and conversely, i.e.,

$$6.4 \quad \text{Dfn} \quad R\,\mathfrak{Isom}_{S,T}(A,B) . \equiv : \mathfrak{Un}_2(R) . \mathfrak{Rel}(R) . \mathfrak{D}(R)=A . \mathfrak{W}(R)=B .$$
$$(u,v)[u\varepsilon A . v\varepsilon A : \supset : uSv . \equiv . (R`u)T(R`v)]$$

If there exists an isomorphism from A to B with respect to S and T, A is called *isomorphic to B with respect to S and T.* If S=T in 6.4 R is said to be an *isomorphism from A to B with respect to S*.

6.41 Dfn R is called an *isomorphism with respect to S* if it is an isomorphism from $\mathfrak{D}(R)$ to $\mathfrak{W}(R)$ with respect to S.

"Isomorphism with respect to an n-adic relation S" is defined accordingly.

The method to be used in constructing the ordinals is due essentially to J. von Neumann. The ordinal α will be the class of all ordinals less than α. For instance, 0 = the null set , $1 = \{0\}$, $2 = \{0,1\}$, ω = the set of all integers , etc. In this way, the class of ordinals will be well ordered by the ε-relation, so that $\alpha\,\varepsilon\,\beta$ corresponds to $\alpha<\beta$. Any ordinal will itself be well ordered by the ε-relation, since an ordinal is a class of ordinals. Moreover, any element of an ordinal must be identical with the segment generated by itself, since this segment is the set of all smaller ordinals. These considerations lead to the following definition:

Definition: X is an *ordinal* if

1. $X\,\mathfrak{We}\,E$,
2. $u\varepsilon X : \supset . u=\mathfrak{Seg}_E(X,u)$.

However, as shown by R. M. Robinson [Journ. of Symb. Log. 2, p. 35. Bernays showed previously that transitivity of E in X and 2' are sufficient.] conditions 1 and 2 may be replaced owing to axiom D by the weaker conditions:

1'. $E\ \mathfrak{Con}\ X$,
2'. $u \varepsilon X . \supset . u \subseteq X$.

X is said to be <u>complete</u> if it has the property 2', i.e. if any element of an element of X is an element of X, that is,

6.5 Dfn $\mathfrak{Comp}(X) . \equiv . (u)[u \varepsilon X . \supset . u \subseteq X]$.
6.51 $\mathfrak{Comp}(X) . \equiv . \mathfrak{S}(X) \subseteq X$

The proof is immediate from 6.5 and 4.8.

6.6 Dfn $\mathfrak{Ord}(X) . \equiv : \mathfrak{Comp}(X) . E\ \mathfrak{Con}\ X$

This definition combines conditions 1' and 2'. An ordinal which is a set is called an <u>ordinal number</u>, denoted by $\mathfrak{O}(X)$.

6.61 Dfn $\mathfrak{O}(X) . \equiv : \mathfrak{Ord}(X) . \mathfrak{M}(X)$

The class of ordinal numbers is denoted by On. [Concerning the normality of $\mathfrak{Ord}$ cf. p.62 .]

6.62 Dfn $x \varepsilon On . \equiv . \mathfrak{O}(x)$
Dfn The letters α, β, γ, ... , will be used to denote variables whose range is the class of ordinal numbers. Evidently these variables are normal.

6.63 Dfn $X < Y . \equiv . X \varepsilon Y$
6.64 Dfn $X \leq Y . \equiv : X < Y . \vee . X = Y$
6.65 $\mathfrak{Comp}(X) . \mathfrak{Comp}(Y) : \supset : \mathfrak{Comp}(X+Y) . \mathfrak{Comp}(X \cdot Y)$

Proof: By 4.88, $\mathfrak{S}(X+Y) = \mathfrak{S}(X) + \mathfrak{S}(Y)$. Therefore, by 6.51, $\mathfrak{Comp}(X+Y)$. Similarly for $X \cdot Y$ by 4.89.

The next step is to show that the definition 6.6 is equivalent to the stronger definition, i.e.

6.7 1. $\mathfrak{Ord}(X) \supset X\ \mathfrak{We}\ E$,
2. $\mathfrak{Ord}(X) . u \varepsilon X \supset u = \mathfrak{Seg}_E(X, u)$.

Proof of 1: Given any non-void subset Y of X, there exists u , by axiom D, such that $u \varepsilon Y$ and $Y \cdot u = 0$, that is, $Y \cdot E``\{u\} = 0$ , since $u = \mathfrak{S}[\{u\}] = E``\{u\}$ by 4.83, 4.82. Therefore $X\ \mathfrak{We}\ E$, by definition 6.2, since $E\ \mathfrak{Con}\ X$, by definition of $\mathfrak{Ord}$. Proof of 2: If $\mathfrak{Ord}(X)$ and $u \varepsilon X$ then $\mathfrak{Seg}_E(X, u) = X \cdot E``\{u\} = X \cdot u = u$, by definition 6.31 and the completeness of X.

7.1 $\mathfrak{Ord}(X).Y\subset X:\supset:\mathfrak{Comp}(Y).\supset.Y\varepsilon X$

Proof: $\mathfrak{S}(Y)\subseteq Y$, so that $E``Y\subseteq Y$, by 4.83. Therefore, by definition 6.3, Y is a section of X. Hence by 6.33 Y must be a segment of X, generated by some element u of X. But then Y=u, by 6.7, hence $Y\varepsilon X$.

7.11 $\mathfrak{Ord}(X).\mathfrak{Ord}(Y):\supset:Y\subset X.\equiv.Y\varepsilon X$

Proof: Since Y is an ordinal, it is complete. Therefore 7.1 establishes one half of the equivalence. The other half merely expresses the fact that X is complete, since Y=X is excluded by 1.6.

7.12 If X and Y are ordinals, one and only one of the following relations holds:

$X\varepsilon Y,\ X=Y,\ Y\varepsilon X.$

Proof: $X\cdot Y\subseteq X$ and $X\cdot Y\subseteq Y$. Suppose now that $X\cdot Y\subset X$ and $X\cdot Y\subset Y$, then $X\cdot Y\varepsilon X$ and $X\cdot Y\varepsilon Y$, by 7.1, since the intersection of two complete classes is complete (6.65). But this implies that $X\cdot Y\varepsilon X\cdot Y$, which is impossible, by 1.6 and axiom A2. Therefore either $X\cdot Y=X$ or $X\cdot Y=Y$, i.e., either $Y\subseteq X$ or $X\subseteq Y$, i.e. $X\subset Y.\vee.X=Y.\vee.Y\subset X$, hence $X\varepsilon Y.\vee.X=Y.\vee.Y\varepsilon X$ by 7.11. Therefore at least one of the three relations holds. Moreover no two can hold simultaneously, since $X\varepsilon X$ or $X\varepsilon Y.Y\varepsilon X$ are impossible, by 1.6 and 1.7 and axiom A2.

7.12 and 6.63 express the fact that any two ordinals are comparable. By 6.1, this implies the statement:

7.13 $E\ \mathfrak{Con}\ On$.

7.14 $\mathfrak{Ord}(A).\supset.A\subseteq On$

Proof: Let A be an ordinal and x an element of A. We have to show that $E\ \mathfrak{Con}\ x$ and $\mathfrak{Comp}(x)$. Take $z\varepsilon y$, $y\varepsilon x$, then since A is complete, $y\varepsilon A$, then iterating, $z\varepsilon A$. E is a relation of well ordering for A therefore transitive in A by 6.21, so that $z\varepsilon x$. Therefore x is complete. $E\ \mathfrak{Con}\ A$ and $x\subseteq A$, so that $E\ \mathfrak{Con}\ x$.

7.15 $\mathfrak{Comp}(On)$

Proof: By 7.14, $x\varepsilon On.\supset.x\subseteq On$.

7.16 $\mathfrak{Ord}(On)$

Proof: 7.13, 7.15, 6.6.

7.161 On (and therefore any class of ordinal numbers) is well ordered by E.

This follows immediately from 7.16 and 6.7 and allows us to prove properties of ordinal numbers by transfinite induction, if the property under consideration is defined by a normal propositional function, since under this assumption the class of ordinal numbers not having the property exists by M2 and (if not empty) contains a smallest element by 7.161 and definition 6.2. By an inductive proof is always meant the reductio ad absurdum of the existence of a smallest ordinal not having the property under consideration.

By 7.14, any element of an ordinal number is itself an ordinal number, so that an ordinal number x is identical with the set of ordinals less than x, recalling that the ε-relation is the ordering relation for ordinals.

7.17 $\mathfrak{Pr}(On)$

Proof: On is an ordinal, so that On would be an ordinal number, if $\mathfrak{M}(On)$, hence $On \varepsilon On$, which is impossible (1.6).

7.2 $\mathfrak{Ord}(X) . \supset : X \varepsilon On . \vee . X = On$. The only ordinal not an ordinal number is On.

Proof: By 7.14, $X \subseteq On$. If $X \subset On$, by 7.11, $X \varepsilon On$.

7.21 Any E-section of an ordinal is an ordinal.

Proof: Any proper E-section of an ordinal X is (by 6.33 and 6.7(2)) an element of X, hence an ordinal by 7.14. A non-proper E-section of X is identical with X.

7.3 $A \subseteq On . \supset . \mathfrak{Ord}[\mathfrak{S}(A)]$

Proof: $\mathfrak{S}(A)$ is complete since, if $x \varepsilon \mathfrak{S}(A)$, there is an ordinal α such that $x \varepsilon \alpha \varepsilon A$, then if $y \varepsilon x$, $y \varepsilon \alpha$, since α is complete, that is $y \varepsilon \mathfrak{S}(A)$. Also $E\, \mathfrak{Con}\, \mathfrak{S}(A)$, for take $x \neq y$, elements of $\mathfrak{S}(A)$, then $x \varepsilon \alpha \varepsilon A$, $y \varepsilon \beta \varepsilon A$. α and β are comparable so that either $\alpha \subseteq \beta$ or $\beta \subseteq \alpha$. Then both x and y are members of the larger of α and β, so that $x \varepsilon y$ or $y \varepsilon x$, since $E\, \mathfrak{Con}\, \alpha$ and $E\, \mathfrak{Con}\, \beta$, that is $E\, \mathfrak{Con}\, \mathfrak{S}(A)$. Therefore $\mathfrak{Ord}[\mathfrak{S}(A)]$.

$\mathfrak{S}(A)$ is the smallest ordinal greater than or equal to all elements of A, i.e. is either the <u>maximum</u> or the <u>limit</u> of the ordinals of A according as to whether there is or is not a greatest ordinal in A. Therefore we use "$\mathfrak{Lim}$" and "$\mathfrak{Max}$" to denote the same operation as $\mathfrak{S}$.

7.31 Dfn $\mathfrak{Lim}(A) = \mathfrak{S}(A)$
$\mathfrak{Max}(A) = \mathfrak{S}(A)$

7.4 Dfn $x\dot{+}1=x+\{x\}$

This defines the successor relation for ordinal numbers as seen by theorems 7.41, 7.411.

7.41 $x\dot{+}1 \varepsilon On. \equiv .x \varepsilon On$

This is easily proved.

7.411 $\sim (\exists\beta)[\alpha<\beta<\alpha\dot{+}1]$

Proof: Suppose $\alpha<\beta<\alpha\dot{+}1$, then $\beta\varepsilon\alpha\dot{+}1$, that is $\beta\varepsilon\alpha+\{\alpha\}$, that is $\beta\varepsilon\alpha$ or $\beta=\alpha$, so that $\beta\leq\alpha$.

Ordinal numbers are to be classified into ordinal numbers of the first kind and ordinal numbers of the second kind as follows:

7.42 Dfn $x\varepsilon K_I. \equiv :(\exists\alpha)[x=\alpha\dot{+}1].\vee.x=0$.

x is of the first kind if it is the successor of an ordinal number or 0. Otherwise x is of the second kind.

7.43 Dfn $K_{II}=On-K_I$
7.44 Dfn $1=0\dot{+}1$
7.45 Dfn $2=1\dot{+}1$

Likewise $3=2\dot{+}1$, etc. Evidently we have:

7.451 If m is a set of ordinal numbers, the ordinal $\alpha=\mathfrak{S}(m)\dot{+}1$ is an ordinal number greater than any element of m .

It will now be shown that it is possible to define functions over On by means of transfinite induction, i.e. determining $F'\alpha$ by means of the behavior of F for ordinal numbers less than α. Since α is the class of ordinals less than α, $F\restriction\alpha$ is F confined to arguments less than α. Therefore the induction should have the form $F'\alpha=G'(F\restriction\alpha)$, where G is a known function. The following theorem, then is what is needed:

7.5 $(G)(E!F)[F\,\mathfrak{Fn}\,On.(\alpha)(F'\alpha=G'(F\restriction\alpha))]$.

Proof: Let us construct F. First, by the existence theorem M2, there exists a class K such that:

$f\varepsilon K.\equiv .(\exists\beta)[f\,\mathfrak{Fn}\,\beta.(\alpha)[\alpha\varepsilon\beta.\supset.f'\alpha=G'(f\restriction\alpha)]]$.

Now set $F=\mathfrak{S}(K)$. If $f,g\varepsilon K$, where $f\,\mathfrak{Fn}\,\beta.g\,\mathfrak{Fn}\,\gamma.\beta\leq\gamma$ it follows that $f=g\restriction\beta$ because for $\alpha\varepsilon\beta$ both f and g satisfy $f'\alpha=G'(f\restriction\alpha)$ (*) and this equation determines an f over β uniquely, as is seen by an induction on α. This means that any two $f,g\varepsilon K$ coincide within the common part of their domains. Therefore F will be a function and its domain will be the sum of the domains of all $f \varepsilon K$ (i.e. $\mathfrak{D}(F)=\mathfrak{S}(Do``K)$) and F will coincide with each $f \varepsilon K$ within $\mathfrak{D}(f)$. F will satisfy (*) for each $\alpha \varepsilon \mathfrak{D}(F)$, because $\alpha \varepsilon \mathfrak{D}(F)$ implies $\alpha \varepsilon \mathfrak{D}(f) \varepsilon On$ for some $f \varepsilon K$ where f satisfies (*) in $\mathfrak{D}(f)$ and $f = F \restriction \mathfrak{D}(f)$. $\mathfrak{D}(F)$ is an ordinal by 7.3 but cannot be an ordinal number α because otherwise F could be extended to a function H over $\alpha \dotplus 1$, by virtue of (*) and M6. But then $\mathfrak{M}(H)$, by 5.19, hence $H \varepsilon K$, which would imply $\alpha \dotplus 1 \subseteq \alpha$. The unicity of F follows by an induction on α.

7.6 Dfn An <u>ordinal function</u> is a function G over an ordinal, with ordinal numbers as values, that is $G \,\mathfrak{Fn}\, \alpha$ (for some α) or $G \,\mathfrak{Fn}\, On$, and $\mathfrak{W}(G) \subseteq On$.

7.61 Dfn An ordinal function G is said to be <u>strictly monotonic</u> if $\alpha < \beta . \supset . G'\alpha < G'\beta$ for $\alpha, \beta \varepsilon \mathfrak{D}(G)$.

By induction it follows that:

7.611 If G is strictly monotonic, then $G`\alpha \geq \alpha$ for $\alpha \varepsilon \mathfrak{D}(G)$.

From this it follows that no two different ordinals X and Y can be isomorphic with respect to E,

7.62 $\mathfrak{Ord}(X) . \mathfrak{Ord}(Y) . H \,\mathfrak{Isom}_{EE}(XY) : \supset : X=Y \cdot H = I \restriction X$.

Proof: By definition of an isomorphism we have: if α, β are elements of X such that $\alpha \varepsilon \beta$, then $H`\alpha \varepsilon H`\beta$, that is, H is strictly monotonic, so that by 7.611 $H`\alpha \geq \alpha$ for $\alpha \varepsilon X$. Likewise, $\breve{H}`(H`\alpha) \geq H`\alpha$, that is, $\alpha \geq H`\alpha$ for $\alpha \varepsilon X$; it follows that $H`\alpha = \alpha$ for $\alpha \varepsilon X$, in other words, X=Y, and $H = I \restriction X$.

As a consequence of 7.62, a well-ordered class can be isomorphic to at most one ordinal. Sufficient conditions for a well-ordered class to be isomorphic to an ordinal are given by the following theorem.

7.7 1. If $\mathfrak{Pr}(A)$ and $A \,\mathfrak{We}\, W$, and if any proper W-section of A is a set, then A is isomorphic to On with respect to W and E.

2. If $a \,\mathfrak{We}\, W$, a is isomorphic to an ordinal number with respect to W and E.

Proof of 1: Let $F`\alpha$ be defined by induction as the first element of A which has not yet occurred as a value of F, that is $F`\alpha$ = first element of $A - \mathfrak{W}(F \restriction \alpha)$. In order to prove the existence of F by 7.5 this condition must be expressed in the form $F`\alpha = G`(F \restriction \alpha)$ (*). Define G by the condition:

$\langle yx \rangle \varepsilon G . \equiv : y \varepsilon (A - \mathfrak{W}(x)) . (A - \mathfrak{W}(x)) . W``\{y\} = 0$,

and define F by (*) and the condition $F \,\mathfrak{Fn}\, On$. Then $G`x \varepsilon A - \mathfrak{W}(x)$ for any set x because $A - \mathfrak{W}(x)$ is a proper class by 5.45, 5.16, hence $\neq 0$. Therefore $F`\alpha\,\varepsilon\,A-\mathfrak{W}(F\restriction\alpha)$ for any α by (*), hence $\mathfrak{W}(F)\subseteq A$. Moreover F is one to one, so that $\mathfrak{W}(F)$, being a one to one image of the proper class On, is itself a proper class by 5.44. But $\mathfrak{W}(F)$ is a section of A, hence by the hypothesis cannot be a proper section, i.e. $\mathfrak{W}(F)=A$. In addition, it is easily seen that $\alpha<\beta .\equiv .(F`\alpha)W(F`\beta)$.

Proof of 2: Construct G and F exactly as in the proof of 1, replacing A by a. Now it can be shown that $a-\mathfrak{W}(F\restriction\alpha)=0$ for some α. In fact, suppose that $(\alpha)[a-\mathfrak{W}(F\restriction\alpha)\neq 0]$, then we could conclude as before, that $\mathfrak{W}(F)\subseteq a$; then $\mathfrak{W}(F)$ would be, as before, a proper class, but this is impossible, since a is a set. Therefore $(\exists\alpha)[a-\mathfrak{W}(F\restriction\alpha)=0]$. Then if α is the smallest ordinal of this kind $F\restriction\alpha$ establishes the isomorphism between a and α. From the axiom of choice it follows:

*7.71 For any set a there exists an ordinal number α and a one to one function g over α such that $a=g``\alpha$.

Proof: By axiom E, the axiom of choice, there is a function C over V such that $x\neq 0 .\supset . C`x\,\varepsilon\,x$. Define F by the postulate $(\alpha)[F`\alpha=C`(a-\mathfrak{W}(F\restriction\alpha))]$ and $F\,\mathfrak{Fn}\,On$. Existence and unicity of F follow from 7.5, if first G is defined by $G`x=C`(a-\mathfrak{W}(x))$, $G\,\mathfrak{Fn}\,V$, using M5. As in the second part of 7.7, it is shown that there exists an α such that $(a-\mathfrak{W}(F\restriction\alpha))=0$. Then if α is the smallest ordinal of this kind, $F\restriction\alpha$ can be taken as g.

It is desirable to assign a well-ordering for the ordered pairs of ordinal numbers:

7.8 Dfn $\langle\alpha\beta\rangle\,Le\,\langle\gamma\delta\rangle .\equiv :\beta<\delta .\vee .(\beta=\delta .\alpha<\gamma) .: Le\subseteq(On^2)^2$,

7.81 Dfn $\langle\alpha\beta\rangle\,R\,\langle\gamma\delta\rangle .$
$$\equiv : \mathfrak{Max}\{\alpha,\beta\}<\mathfrak{Max}\{\gamma,\delta\} .\vee .[\mathfrak{Max}\{\alpha,\beta\}=\mathfrak{Max}\{\gamma,\delta\} .\langle\alpha\beta\rangle\,Le\,\langle\gamma\delta\rangle] .: R\subseteq(On^2)^2 .$$

The existence of an Le satisfying 7.8 follows from M4 since the relation Le defined by:
$$\langle xy\rangle\,\varepsilon\,Le \equiv (\exists\alpha\beta\gamma\delta)[x=\langle\alpha\beta\rangle .y=\langle\gamma\delta\rangle :\beta<\delta .\vee .(\beta=\delta .\alpha<\gamma)]$$
evidently satisfies 7.8. Similarly for R. On^2 is well ordered by R in such a way that:

7.811 Any proper R-section of On^2 is a set.

Proof: Consider a pair $\langle\mu\nu\rangle$ such that $\langle\mu\nu\rangle\,R\,\langle\alpha\beta\rangle$, then:
$$\mathfrak{Max}\{\mu\nu\}\leq\mathfrak{Max}\{\alpha\beta\}<\mathfrak{Max}\{\alpha\beta\}+1 .$$
Therefore $\mu,\nu\,\varepsilon\,[\mathfrak{Max}\{\alpha\beta\}+1]$, so that $\langle\mu\nu\rangle\varepsilon a$, where $a=[\mathfrak{Max}\{\alpha\beta\}+1]^2$. a is a set by 5.18. Therefore the class of all pairs $\langle\mu\nu\rangle$ such that $\langle\mu\nu\rangle\,R\,\langle\alpha\beta\rangle$ is contained in the set a, hence is itself a set.

Now, applying 7.7, (since On^2 is a proper class by 7.17, 5.43), we have:

7.82 On^2 is isomorphic to On, with respect to R and E. Let the isomorphism from On^2 to On be denoted by P, i.e.

7.9 Dfn $P\,\mathfrak{Fn}\,On^2 . \mathfrak{W}(P)$
$= On : (\alpha, \beta, \gamma, \delta)[\langle\alpha\beta\rangle\, R \langle\gamma\delta\rangle . \supset . P`\langle\alpha\beta\rangle < P`\langle\gamma\delta\rangle]$.

7.91 $P`\langle\alpha\beta\rangle \geq \mathfrak{Max}\{\alpha\beta\}$

Proof: Take $\gamma = \mathfrak{Max}\{\alpha\beta\}$. Then $P`\langle\alpha\beta\rangle \geq P`\langle\gamma o\rangle$ by 7.9 but since $P`\langle\gamma o\rangle$ considered as a function of $\gamma$ is strictly monotonic by 7.9 we have $\gamma \leq P`\langle\gamma o\rangle$ by 7.611, i.e., $P`\langle\alpha\beta\rangle \geq \mathfrak{Max}\{\alpha\beta\}$.

CHAPTER IV

CARDINAL NUMBERS

We can now proceed with the theory of cardinals. Most of the theorems and definitions of this chapter (except those concerning finite cardinals) depend in our development on the axiom of choice, although its use could be avoided in many cases. Two classes X and Y are said to be equivalent if there is a one to one correspondence between the elements of each, i.e.,

8.1 Dfn $X \simeq Y . \equiv . (\exists Z)[\mathfrak{Un}_2(Z) . \mathfrak{Rel}(Z) . \mathfrak{D}(Z)=X . \mathfrak{W}(Z)=Y]$.

This notion is not normal★; the corresponding normal notion is as follows:

8.12 Dfn $X \simeq' Y . \equiv . (\exists z)[\mathfrak{Un}_2(z) . \mathfrak{Rel}(z) . \mathfrak{D}(z)=X . \mathfrak{W}(z)=Y]$.

8.121 $x \simeq y . \equiv . x \simeq' y$

Proof: A class Z satisfying the right hand side of 8.1 for two sets X, Y is a set by 5.19.

8.13 Dfn $\langle xy\rangle \varepsilon Aeq . \equiv . x \simeq y : \mathfrak{Rel}(Aeq)$

The <u>cardinal of X</u>, denoted by $\bar{\bar{X}}$, is defined by the postulate:◇

*8.2 Dfn $x \simeq \bar{\bar{x}} . \bar{\bar{x}} \varepsilon On . (\alpha)[\alpha < \bar{\bar{x}} . \supset . \sim(\alpha \simeq x)] . \mathfrak{Pr}(X) . \supset . \bar{\bar{X}} = On$

By theorem 7.71 it is seen that $\bar{\bar{X}}$ exists. The unicity is immediate. $\bar{\bar{X}}$ is a normal operation, since

$X \varepsilon \bar{\bar{Y}} . \equiv : X \varepsilon On . (\alpha)[\alpha \simeq' Y . \supset . X \varepsilon \alpha]$.

Hence by M5 there exists a function Nc over V such that $Nc'x = \bar{\bar{x}}$ for any set x .

*8.20 Dfn $Nc'x = \bar{\bar{x}} . Nc \, \mathfrak{Fn} \, V$

The cardinal of a set is called a <u>cardinal number</u>, i.e., the class N of cardinal numbers is defined by:

*8.21 Dfn $N = \mathfrak{W}(Nc)$.

*8.22 $N \subseteq On$

This follows immediately from 8.2, 8.21.

★ See Note 8
◇ See Note 9 } of Notes added to the second printing, pp. 68-9.

An ordinal number is a cardinal number if and only if it is equivalent to no smaller ordinal, i.e., if it is an initial number in the usual terminology.[7]

The next five theorems are immediate consequences of the definition of cardinals.

*8.23 $x \varepsilon N . \equiv . x = \bar{\bar{x}}$
*8.24 $\bar{\bar{x}} \simeq x$
*8.25 $x \simeq y . \equiv . \bar{\bar{x}} = \bar{\bar{y}}$
*8.26 $\bar{\bar{\alpha}} \le \alpha$
*8.27 $Nc`[Nc`x] = Nc`x$
*8.28 $x \subseteq y . \supset . \bar{\bar{x}} \le \bar{\bar{y}}$

Proof: $y \simeq \bar{\bar{y}}$, therefore there exists a $z \subseteq \bar{\bar{y}}$ such that $x \simeq z$. z is a set of ordinal numbers, hence well-ordered by E, hence is isomorphic to an ordinal number β by 7.7, i.e. there is an h such that: $h\ \mathfrak{Isom}_{EE}(\beta, z)$. Hence $\alpha \le h`\alpha$ for $\alpha \varepsilon \beta$ by 7.611. But $h`\alpha \varepsilon z \subseteq \bar{\bar{y}}$ for $\alpha \varepsilon \beta$. Hence $\alpha \varepsilon \beta \supset . \alpha \le h`\alpha \varepsilon \bar{\bar{y}}$, that is $\beta \subseteq \bar{\bar{y}}$. But $\bar{\bar{\beta}} = \bar{\bar{z}}$, therefore $\bar{\bar{z}} = \bar{\bar{\beta}} \le \beta \le \bar{\bar{y}}$.

The Schroeder-Bernstein Theorem appears as a consequence. Namely, if $x \simeq t \subseteq y$ and $y \simeq u \subseteq x$, then $\bar{\bar{x}} = \bar{\bar{y}}$, since $\bar{\bar{x}} = \bar{\bar{t}} \le \bar{\bar{y}}$, and $\bar{\bar{y}} = \bar{\bar{u}} \le \bar{\bar{x}}$. This proof depends, however, on the axiom of choice.

The proofs of the next three theorems are omitted.

*8.3 $\alpha \dotplus 1 \le \overline{\overline{\alpha^2}}$ for $\alpha > 1$
*8.31 $\mathfrak{Un}(A) . \supset . \overline{\overline{A``x}} \le \bar{\bar{x}}$
*8.32 $\overline{\overline{\mathfrak{P}(x)}} > \bar{\bar{x}}$ (Cantor's theorem)
*8.33 $\mathfrak{Pr}(N)$

Proof: Take $m \subseteq N$, then by 8.32 $\overline{\overline{\mathfrak{P}(\mathfrak{S}(m))}} > \overline{\overline{\mathfrak{S}(m)}}$. But $\overline{\overline{\mathfrak{S}(m)}} \ge \alpha$, where α is any member of m by 8.28, 8.23. Therefore there is a cardinal number greater than any element of m ; hence $m \neq N$, i.e. N can not be a set.

We now define the class ω of integers:

8.4 Dfn $x \varepsilon \omega . \equiv . x + \{x\} \subseteq K_I$,

i.e. x is an integer if it is an ordinal number of the first kind and if all smaller ordinals are likewise of the first kind. It follows immediately that :

8.41 $\alpha \varepsilon \omega . \supset . \alpha \dotplus 1 \varepsilon \omega$ and $\alpha \varepsilon \omega . \beta < \alpha : \supset . \beta \varepsilon \omega$.

Dfn i, k are variables whose range is ω.

The principle of induction holds for integers:

8.44 $0 \varepsilon A . (k)[k \varepsilon A . \supset . k \dotplus 1 \varepsilon A] : \supset . \omega \subseteq A$.

---

[7] This treatment of cardinals is due to v. Neumann [cf. Math. Z. 27, p. 731].

Proof: If $\omega \subseteq A$ is false there must be a smallest i such that i is not a member of A. This leads to a contradiction with the hypothesis, since either $i=0$ or $i=k\dot{+}1$ by 8.4 and 7.42.

Functions over the class of integers may be defined inductively:

8.45 $(a,G)(E!F)[F\mathfrak{Fn}\,\omega . F`0=a.(k)(F`(k\dot{+}1)=G`(F`k))]$.

This can be proved either by specializing G in 7.5, or by arguments similar to those used in the proof of 7.5.

8.46 $i \neq k . \supset . \sim(i \simeq k)$

This can be proved by induction on integers, since
$i\dot{+}1 \simeq k\dot{+}1 \supset i \simeq k$.

8.461 $\alpha \neq k . \supset . \sim(\alpha \simeq k)$

Proof by induction on k since $k \dot{+} 1 \simeq \alpha \geq \omega$ would imply $k \simeq \alpha$.

*8.47 $i \varepsilon N$

This follows from 8.46.

A class is called finite if it is equivalent to an integer; otherwise infinite, i.e.

8.48 Dfn $\mathfrak{Fin}(x) . \equiv . (\exists \alpha)[\alpha \varepsilon \omega . \alpha \simeq x]$,
8.49 Dfn $\mathfrak{Inf}(x) . \equiv . \sim \mathfrak{Fin}(x)$.
8.491 $\mathfrak{Fin}(x) . z \subseteq x : \supset . \mathfrak{Fin}(z)$
$\mathfrak{Fin}(x) . \mathfrak{Fin}(y) : \supset : \mathfrak{Fin}(x+y) . \mathfrak{Fin}(x \times y)$

This is proved by an induction on the integer i equivalent to x .

8.492 $\mathfrak{Fin}(\alpha) . \equiv . \alpha \varepsilon \omega$

This follows from 8.461.

8.5 $\mathfrak{Ord}(\omega)$

Proof: ω is a class of ordinal numbers, hence $E\,\mathfrak{Con}\,\omega$. Moreover, every element of an integer is an integer by definition 8.4, that is, $\mathfrak{Comp}(\omega)$. Therefore $\mathfrak{Ord}(\omega)$.

8.51 $\mathfrak{M}(\omega)$

Proof: Axiom C1 (the axiom of infinity) provides for the existence of a non-empty set b , such that for every $x \varepsilon b$ there is a $y \varepsilon b$ which contains exactly one element more than x ; namely take for b the class of all subsets of elements of the set a, whose existence is postulated by axiom C1 (b is a set because $b \subseteq \mathfrak{P}[\mathfrak{S}(a)]$). Now consider the class c defined by $c = (\omega 1 Aeq)``b$ i.e. the class of integers equivalent to elements of b . c is a set by 5.1 and 8.46, and $\omega \subseteq c$, as can be shown by induction owing to the above mentioned property of b .

8.52 $\omega \varepsilon K_{II}$

Proof: $x \varepsilon \omega . \supset . x+1 \varepsilon \omega$, by 8.41. If $\omega = \alpha+1$, we would have $\alpha \varepsilon \omega$ by 7.4, hence $\alpha+1 \varepsilon \omega$ i.e. $\omega \varepsilon \omega$, which is impossible.

8.53 There exists an ordinal number of the second kind.

Proof: By 8.52, ω is such an ordinal.

*8.54 Dfn $N' = N - \omega$

*8.55 $N' \subseteq On$

*8.56 N' is isomorphic to On with respect to E.

Proof: $\mathfrak{Pr}(N')$ by 5.45, since $\mathfrak{M}(\omega)$. Moreover any proper section of N' is generated by an $\alpha \varepsilon N'$, hence $\subseteq \alpha$, hence a set. Therefore 7.7 gives the result. The isomorphism from On to N' is denoted by $\aleph$, i.e.

*8.57 Dfn $\aleph\, \mathfrak{Isom}_{EE}(On, N')$.

It follows:

*8.58 $\aleph`0 = \omega$

since $\omega \varepsilon N$ by 8.461. $\aleph_\gamma$ and ω_γ are defined by:

*8.59 Dfn $\aleph_\gamma = \omega_\gamma = \aleph`\gamma$.

*8.62 $\overline{\overline{\aleph_\alpha^2}} = \aleph_\alpha$

Proof: Assuming γ to be the smallest ordinal number for which $\aleph_\gamma^2 \neq \aleph_\gamma$, we prove $\omega_\gamma^2 \simeq \omega_\gamma$. To this end, owing to the Schroeder-Bernstein Theorem it is sufficient to show $P``(\omega_\gamma^2) \subseteq \omega_\gamma$, i.e. $P`\langle\alpha\beta\rangle < \omega_\gamma$ for $\alpha, \beta < \omega_\gamma$, where P is the function defined by 7.9. Since, for every δ, $\delta < \omega_\gamma . \equiv . \overline{\overline{\delta}} < \overline{\overline{\omega_\gamma}}$, it is sufficient to show: $\overline{\overline{P`\langle\alpha\beta\rangle}} < \omega_\gamma$ for $\alpha, \beta < \omega_\gamma$ . Now $\overline{\overline{P`\langle\alpha\beta\rangle}}$ is the power of the set of ordinals $< P`\langle\alpha\beta\rangle$ . This set by definition of P (7.9) is mapped by P on the set m of pairs preceding $\langle\alpha\beta\rangle$ in the ordering R. Hence $\overline{\overline{P`\langle\alpha\beta\rangle}} = \overline{\overline{m}}$, but (as seen in the proof of 7.811) $m \subseteq (\mu+1)^2$ where $\mu = \mathfrak{Max}\{\alpha\beta\}$.

Now we distinguish two cases:

1. μ is finite: then $(\mu+1)^2<\omega$ by 8.491. Hence $\overline{\overline{m}}\leq(\mu+1)^2<\omega\leq\omega_\gamma$ in this case.

2. μ is infinite: then, since $\mu<\omega_\gamma$ by assumption, $\overline{\overline{\mu}}=\omega_\delta$ for some $\delta<\gamma$. Hence $\overline{\overline{\mu^2}}=\overline{\overline{\mu}}$ by the inductive assumption. Hence (using *8.3) $\overline{\overline{m}}\leq\overline{\overline{(\mu+1)^2}}\leq\overline{\overline{(\mu^2)^2}}=\overline{\overline{\mu}}<\omega_\gamma$ also in this case.

It results that:

*8.621 For any infinite set x, $\overline{\overline{x^2}}=\overline{\overline{x}}$,

and therefore

*8.63 $\mathfrak{Inf}(x).y\neq 0:\supset:\overline{\overline{x\times y}}=\overline{\overline{x+y}}=\mathfrak{Max}(\overline{\overline{x}},\overline{\overline{y}})$.

Furthermore

*8.64 If for any $y\,\varepsilon\,m$, $\overline{\overline{F'y}}\leq\overline{\overline{a}}$, then $\overline{\overline{\mathfrak{S}(F''m)}}\leq\overline{\overline{a\times m}}$.

The proofs of these results on cardinals are not included, since they do not differ from the usual proofs.

8.7 Dfn A is closed with respect to R if $R''A\subseteq A$.

8.71 Dfn A is closed with respect to S as a triadic relation if $S''(A^2)\subseteq A$.

8.72 Y is called closure of X with respect to $R_1,\dots,R_k$ and with respect to $S_1,\dots,S_j$ as triadic relations if Y is the smallest class including X which is closed with respect to the R's and closed with respect to the S's as triadic relations.

The existence of this class will be needed only under the following conditions:

*8.73 If $\mathfrak{M}(X)$ and if the R's and S's are single-valued, then the closure Y exists and is a set, and if in addition X is infinite then $\overline{\overline{Y}}=\overline{\overline{X}}$.

Proof: Define $G\,\mathfrak{Fn}\,V$ as follows:

$$G'x=x+R_1''x+\dots+R_k''x+S_1''(x^2)+\dots+S_j''(x^2).$$

The right-hand side is normal, and by 5.1, 5.13, 5.18 is a set for any set x, hence G exists by M5. Now define $f\,\mathfrak{Fn}\,\omega$ by 8.45 as follows:

$$f'0=x,\quad f'(k+1)=G'f'(k).$$

Now consider $\mathfrak{S}(f''\omega)$; this is a set, and satisfies the requirements of definition 8.72. Now for any infinite set y we have $\overline{\overline{G'y}}=\overline{\overline{y}}$ by 8.31, 8.621, 8.63. Therefore, if x is infinite, $\overline{\overline{f'n}}=\overline{\overline{f'0}}=\overline{\overline{x}}$, by complete induction on n. Hence

$\overline{\overline{\mathfrak{S}(f''\omega)}}\leq\overline{\overline{x\times\omega}}=\mathfrak{Max}(\overline{\overline{x}},\overline{\overline{\omega}})=\overline{\overline{x}}$ by 8.64, 8.63 and

$\overline{\overline{\mathfrak{S}(f''\omega)}}\geq\overline{\overline{f'0}}=\overline{\overline{x}}$ by 8.28.

CHAPTER V

THE MODEL Δ

The classes and sets of the model Δ will form a certain subfamily of the classes and sets of our original system Σ and the ε-relation of the model Δ will be the original ε-relation confined to the classes and sets of Δ. We call the classes and sets of Δ <u>constructible</u> and denote the notion of constructible class by $\mathfrak{L}$ and the class of constructible sets by L. Constructible sets are those which can be obtained by iterated application of the operations given by axioms A4, B1-8, modified so that they yield sets if applied to sets. In addition, at certain stages of this generating process the set of all previously obtained sets will be added as a new constructible set. This permits the generating process to continue into the transfinite. The above mentioned axioms lead to the following eight binary operations $\mathfrak{F}_1,\dots,\mathfrak{F}_8$ called fundamental operations:

9.1 Dfn $\mathfrak{F}_1(XY)=\{XY\}$,
$\mathfrak{F}_2(XY)=E\cdot X$,
$\mathfrak{F}_3(XY)=X-Y$,
$\mathfrak{F}_4(XY)=X\restriction Y$ [i.e. $=X\cdot(V\times Y)$] ,
$\mathfrak{F}_5(XY)=X\cdot\mathfrak{D}(Y)$,
$\mathfrak{F}_6(XY)=X\cdot Y^{-1}$,
$\mathfrak{F}_7(XY)=X\cdot\,\mathfrak{Cnv}_2(Y)$,
$\mathfrak{F}_8(XY)=X\cdot\,\mathfrak{Cnv}_3(Y)$.

The factor X in $\mathfrak{F}_2,\dots,\mathfrak{F}_8$ is added for reasons that will appear later (theorem 9.5). The operation of intersection (given by axiom B2) is left out because $X\cdot Y=X-(X-Y)$. Owing to 4.92-4.96, $\mathfrak{F}_4,\dots,\mathfrak{F}_8$ can be expressed differently as follows:

9.11 $\mathfrak{F}_4(XY)=X\cdot\breve{P}_2``Y$,
$\mathfrak{F}_5(XY)=X\cdot P_2``Y$,
$\mathfrak{F}_6(XY)=X\cdot P_3``Y$,
$\mathfrak{F}_7(XY)=X\cdot P_4``Y$,
$\mathfrak{F}_8(XY)=X\cdot P_5``Y$.

In other words,

9.12 $\mathfrak{F}_i(XY)=X\cdot Q_i``Y$; $i=4,\dots,8$,

where the Q_i are defined by

9.14 $Q_4 = \breve{P}_2$, $Q_5 = P_2$, $Q_6 = P_3$, $Q_7 = P_4$, $Q_8 = P_5$.

By means of theorem 5.11 it is seen that all the fundamental operations give sets when applied to sets.

Now consider the class $9 \times On^2$ (i.e. the class of triples $\{i\alpha\beta\}$, $i<9$) and define the following well-ordering relation S for it:

9.2 Dfn $\mu,\nu<9 . \supset .:$
$\langle\mu\alpha\beta\rangle S \langle\nu\gamma\delta\rangle . \equiv : \langle\alpha\beta\rangle R \langle\gamma\delta\rangle . \vee . (\langle\alpha\beta\rangle = \langle\gamma\delta\rangle . \mu<\nu) :: S \subseteq (9\times On^2)^2$,

where R is the relation defined by 7.81. Concerning the existence of S cf. definition 7.8. Since

$$\langle i\alpha\beta\rangle S \langle j\gamma\delta\rangle . \supset : \langle\alpha\beta\rangle R \langle\gamma\delta\rangle . \vee . \langle\alpha\beta\rangle = \langle\gamma\delta\rangle$$

it follows from 7.811 and 5.18 that any proper S-section of $9\times On^2$ is a set. But $9\times On^2$ is not a set by 5.43. Hence $9\times On^2$ is isomorphic to On with respect to S and E by 7.7, i.e. there exists a J satisfying the following defining postulate:

9.21 Dfn $J \,\mathfrak{Fn}(9\times On^2) . \mathfrak{W}(J) = On :$
$\mu,\nu<9 . \supset [\langle\mu\alpha\beta\rangle S \langle\nu\gamma\delta\rangle . \supset . J‘\langle\mu\alpha\beta\rangle < J‘\langle\nu\gamma\delta\rangle]$.

Now we define nine functions $J_0,\ldots,J_8$ over On^2 by:

9.22 Dfn $J_0‘\langle\alpha\beta\rangle = J‘\langle 0\alpha\beta\rangle$, $J_0 \,\mathfrak{Fn}\, On^2$,
.... ,
$J_8‘\langle\alpha\beta\rangle = J‘\langle 8\alpha\beta\rangle$, $J_8 \,\mathfrak{Fn}\, On^2$.

Evidently we have:

9.23 The $\mathfrak{W}(J_i)$, $i=0,\ldots,8$ are mutually exclusive and their sum is On. [It is easily seen, but not used in the sequel, that the $\mathfrak{W}(J_i)$ are the congruence classes of On mod. 9 and that $J_i‘\langle\alpha\beta\rangle = 9 \dot\times P‘\langle\alpha\beta\rangle \dot+ i$, where $\dot+$ and $\times$ denote arithmetic addition and multiplication of ordinals.]

By definition of J there exists for any γ a unique triple $\langle i\alpha\beta\rangle$ such that $\gamma = J‘\langle i\alpha\beta\rangle$. Hence there are two functions K_1, K_2 over On such that: $K_1‘ J_i‘ \langle\alpha\beta\rangle = \alpha$, $K_2‘ J_i‘\langle\alpha\beta\rangle = \beta$, for any $i<9$. K_1, K_2 are defined by:

9.24 Dfn 1. $\langle\alpha\gamma\rangle \in K_1 \equiv (\exists\mu,\beta)[\mu<9 . \gamma = J‘\langle\mu\alpha\beta\rangle] . K_1 \subseteq On^2$,
$\langle\beta\gamma\rangle \in K_2 \equiv (\exists\mu,\alpha)[\mu<9 . \gamma = J‘\langle\mu\alpha\beta\rangle] . K_2 \subseteq On^2$.

For the J_i and K_i we have the following theorems:

9.25 $J_i‘\langle\alpha\beta\rangle \geq \mathfrak{Max}\{\alpha\beta\}$,
$J_i‘\langle\alpha\beta\rangle > \mathfrak{Max}\{\alpha\beta\}$ for $i \neq 0$,
$K_1‘\alpha \leq \alpha$, $K_2‘\alpha \leq \alpha$,

$$K_1{}^{\text{‘}}\alpha < \alpha \;,\quad K_2{}^{\text{‘}}\alpha < \alpha \quad \text{for} \quad \alpha \notin \mathfrak{W}(J_0) \quad .$$

Proof: Set $\mathfrak{Max}\{\alpha\beta\} = \gamma$: then we have $J_0{}^{\text{‘}}\langle\alpha\beta\rangle \geq J_0{}^{\text{‘}}\langle\gamma 0\rangle$ by definition 9.21; $J_0{}^{\text{‘}}\langle\gamma 0\rangle \geq \gamma$ by 7.611; $J_i{}^{\text{‘}}\langle\alpha\beta\rangle > J_0{}^{\text{‘}}\langle\alpha\beta\rangle$ for $i \neq 0$ by definition 9.21. Writing the last three inequalities as one inequality we obtain (for $i \neq 0$):

$$J_i{}^{\text{‘}}\langle\alpha\beta\rangle > J_0{}^{\text{‘}}\langle\alpha\beta\rangle \geq J_0{}^{\text{‘}}\langle\gamma 0\rangle \geq \gamma \;,$$

which gives the first two statements of 9.25. The last two express the same facts in terms of K_1 and K_2.

*9.26 $\alpha,\beta < \omega_\gamma \supset J_i{}^{\text{‘}}\langle\alpha\beta\rangle < \omega_\gamma$

Proof: By definition 9.21 J maps the set m of triples preceding $\langle i\alpha\beta\rangle$ in the ordering S on the set of ordinals $< J_i{}^{\text{‘}}\langle\alpha\beta\rangle$. Hence $J_i{}^{\text{‘}}\langle\alpha\beta\rangle \simeq m$. But $m \subseteq 9 \times (\mathfrak{Max}\{\alpha\beta\} \dot{+} 1)^2$ by 9.2 and 7.81. Hence the theorem by 8.491 or 8.63 according as $\gamma = 0$ or $\gamma > 0$ (using 8.492 in the first case). Note that the axiom of choice is not used in the case $\gamma = 0$.

*9.27 $\omega_\alpha \,\varepsilon\, \mathfrak{W}(J_0)$

Proof: $\omega_\alpha \leq J^{\text{‘}}\langle 0\,\omega_\alpha 0\rangle$ by 9.25, but not $\omega_\alpha < J^{\text{‘}}\langle 0\,\omega_\alpha 0\rangle$, because this would imply $\omega_\alpha = J^{\text{‘}}\langle i\gamma\delta\rangle$ for some triple $\langle i\gamma\delta\rangle$ preceding $\langle 0\,\omega_\alpha 0\rangle$ in the ordering S. But $\langle i\gamma\delta\rangle \,S\, \langle 0\,\omega_\alpha 0\rangle$ implies $\gamma,\delta < \omega_\alpha$ hence $J^{\text{‘}}\langle i\gamma\delta\rangle < \omega_\alpha$ by 9.26. Hence $\omega_\alpha = J_0{}^{\text{‘}}\langle\omega_\alpha 0\rangle$ i.e. $\omega_\alpha \,\varepsilon\, \mathfrak{W}(J_0)$. For $\alpha = 0$ the axiom of choice is not used in this argument.

Now we define by transfinite induction a function F [the letter F is to be used only as a constant from now on. A similar remark applies to R, S, C defined respectively by 7.81, 9.2, 11.81] over On by the following postulates:

9.3 Dfn
$$\begin{aligned} &\alpha \,\varepsilon\, \mathfrak{W}(J_0) \supset F^{\text{‘}}\alpha = \mathfrak{W}(F \restriction \alpha) \;, \\ &\alpha \,\varepsilon\, \mathfrak{W}(J_1) \supset F^{\text{‘}}\alpha = \mathfrak{F}_1(F^{\text{‘}}K_1{}^{\text{‘}}\alpha, F^{\text{‘}}K_2{}^{\text{‘}}\alpha) \;, \\ &\cdots\cdots\cdots\cdots\cdots\cdots\cdots\cdots \;, \\ &\alpha \,\varepsilon\, \mathfrak{W}(J_8) \supset F^{\text{‘}}\alpha = \mathfrak{F}_8(F^{\text{‘}}K_1{}^{\text{‘}}\alpha, F^{\text{‘}}K_2{}^{\text{‘}}\alpha) \;. \end{aligned}$$

$F\,\mathfrak{Fn}\,On$.

In order to prove the existence of F by 7.5 it is necessary to define first a function G over V by the following postulates: If $\mathfrak{D}(x) \,\varepsilon\, \mathfrak{W}(J_0)$, $G^{\text{‘}}x = \mathfrak{W}(x)$, if $\mathfrak{D}(x) \,\varepsilon\, \mathfrak{W}(J_i), i=1,2,\ldots,8$, $G^{\text{‘}}x = \mathfrak{F}_i[x^{\text{‘}}K_1{}^{\text{‘}}\mathfrak{D}(x), x^{\text{‘}}K_2{}^{\text{‘}}\mathfrak{D}(x)]$ and $G^{\text{‘}}x = 0$ everywhere else. Since all symbols occuring are nórmal [cf. p.62] G exists by M6. By 7.5 there exists an F over On satisfying the equation $F^{\text{‘}}\alpha = G^{\text{‘}}(F \restriction \alpha)$ which implies that F satisfies 9.3 as is seen by the following proof: Suppose $\alpha \,\varepsilon\, \mathfrak{W}(J_i), i \neq 0$. Then, since $\mathfrak{D}(F \restriction \alpha) = \alpha$, $\mathfrak{D}(F \restriction \alpha) \,\varepsilon\, \mathfrak{W}(J_i)$. Therefore

$$G^{\text{‘}}(F \restriction \alpha) = \mathfrak{F}_i[(F \restriction \alpha)^{\text{‘}}K_1{}^{\text{‘}}\alpha, (F \restriction \alpha)^{\text{‘}}K_2{}^{\text{‘}}\alpha]$$

$K_1{}^{\text{‘}}\alpha < \alpha$ and $K_2{}^{\text{‘}}\alpha < \alpha$, by 9.25, and $(F \restriction \alpha)^{\text{‘}}\beta = F^{\text{‘}}\beta$ if $\beta < \alpha$, therefore $F'\alpha = G'(F\restriction\alpha) = \mathfrak{F}_i[F'K_1'\alpha, F'K_2'\alpha]$. Similarly, if $\alpha\,\varepsilon\,\mathfrak{W}(J_0)$, then $\mathfrak{D}(F\restriction\alpha)\,\varepsilon\,\mathfrak{W}(J_0)$, so that $F'\alpha = G'(F\restriction\alpha) = \mathfrak{W}(F\restriction\alpha)$.

Hence F exists and by induction it is seen that F is uniquely determined. The following results are consequences of 9.3 obtained by substituting $J_i'\langle\beta\gamma\rangle$ for α in the ith line of 9.3 and applying the equations: $K_1'\,J_i'\langle\alpha\beta\rangle = \alpha$, $K_2'\,J_i'\langle\alpha\beta\rangle = \beta$ which hold by definition 9.24.

9.31 $F'J_1'\langle\beta\gamma\rangle = \{F'\beta\, F'\gamma\}$

9.32 $F'J_2'\langle\beta\gamma\rangle = E\cdot F'\beta$

9.33 $F'J_3'\langle\beta\gamma\rangle = F'\beta - F'\gamma$

9.34 $F'J_i'\langle\beta\gamma\rangle = F'\beta\cdot Q_i''(F'\gamma)$; $i=4,5,\ldots,8$

9.35 $\alpha\,\varepsilon\,\mathfrak{W}(J_0) . \supset . F'\alpha = F''\alpha$

The last set of theorems shows how F reflects the nine fundamental operations of 9.1.

A set x is said to be <u>constructible</u> if there exists an α such that $x = F'\alpha$. The class of constructible sets is denoted by L, i.e.

9.4 Dfn $L = \mathfrak{W}(F)$.

A class A is constructible if all its elements are constructible sets and if the intersection of A with any constructible set is also a constructible set, i.e.

9.41 Dfn $\mathfrak{L}(A) .\equiv .: A \subseteq L : x\varepsilon L . \supset . A\cdot x\,\varepsilon\, L$.

Dfn $\bar{x},\ldots,\bar{z}$ will be used as variables for constructible sets and $\bar{X},\ldots,\bar{Z}$ as variables for constructible classes.

9.42 Dfn The smallest α such that $x = F'\alpha$ is called the <u>order of x</u> and is denoted by $Od'x$, i.e.

9.421 Dfn $\langle yx\rangle\,\varepsilon\, Od \equiv \langle xy\rangle\,\varepsilon\, F . (z)[z\varepsilon y \supset \sim \langle xz\rangle\,\varepsilon\, F] . Od \subseteq V^2$.

9.5 $\mathfrak{Comp}\,(F''\alpha)$

It is sufficient to prove: $F'\alpha \subseteq F''\alpha$, i.e., all elements of a constructible set appear earlier than the set itself.

Proof: Let α be the first ordinal for which $F'\alpha \subseteq F''\alpha$ is false. If $\alpha\,\varepsilon\,\mathfrak{W}(J_0)$ then $F'\alpha = F''\alpha$ hence $F'\alpha \subseteq F''\alpha$. If $\alpha\,\varepsilon\,\mathfrak{W}(J_i), i\neq 0$, then $\alpha = J_i'\langle\beta\gamma\rangle, i\neq 0$. By theorems 9.32, 9.33, 9.34, if $i>1$, $F'\alpha \subseteq F'\beta$. But $\beta<\alpha$, by 9.25, so that the theorem holds for β , that is, $F'\beta \subseteq F''\beta$. Hence $F'\alpha \subseteq F''\beta$. Again, since $\beta<\alpha$, $F''\beta \subseteq F''\alpha$, therefore $F'\alpha \subseteq F''\alpha$. If $i=1$, by 9.31 $F'\alpha = \{F'\beta\, F'\gamma\}$ where $\alpha = J_1'\langle\beta\gamma\rangle$. By 9.25, $\beta,\gamma<\alpha$. Therefore $F'\beta\,\varepsilon\, F''\alpha$ and $F'\gamma\,\varepsilon\, F''\alpha$, hence $\{F'\beta\, F'\gamma\} \subseteq F''\alpha$, i.e. $F'\alpha \subseteq F''\alpha$.

9.51 $\mathfrak{Comp}(L)'$, i.e. any element of a constructible set is constructible. [For constructible classes the same thing is true by definition 9.41.]

Proof: Take $x \varepsilon L$ and let $\alpha = Od`x$, so that $F`\alpha = x$. Then by 9.5, $x \subseteq F``\alpha$. Hence $x \subseteq L$, since $F``\alpha \subseteq L$.

The following statement follows from 9.5:

9.52 If $x \varepsilon y$, and $x, y \varepsilon L$, then $Od`x < Od`y$. In other words $x \varepsilon F`\alpha . \supset Od`x < \alpha$.

$\mathfrak{F}_1, \ldots, \mathfrak{F}_8$ yield constructible sets if applied to constructible sets, i.e.

9.6 $\mathfrak{F}_i(\overline{xy}) \varepsilon L; \ i = 1, \ldots, 8$.

Proof: There exist β, γ such that $\bar{x} = F`\beta$, $\bar{y} = F`\gamma$; 9.31 to 9.34 give the result.

9.61 $\bar{x} \cdot \bar{y} \varepsilon L$

Proof: $\bar{x} \cdot \bar{y} = \bar{x} - (\bar{x} - \bar{y})$, then 9.6 for $i = 3$ gives the theorem.

9.611 $Od`\bar{x} < \omega_\alpha . Od`\bar{y} < \omega_\alpha : \supset . Od`(\bar{x} \cdot \bar{y}) < \omega_\alpha$

Proof by 9.26.

9.62 $x, y \varepsilon L . \equiv . \langle xy \rangle \varepsilon L$ and $x . y . z \varepsilon L . \equiv . \langle xyz \rangle \varepsilon L$

Proof: The implication in one direction results from expressing $\langle xy \rangle$ as $\{\{x\}\{xy\}\}$, then applying 9.6, and the reverse implication is a consequence of 9 51.

9.621 $\langle xy \rangle \varepsilon L . \equiv . \langle yx \rangle \varepsilon L$
$\langle xyz \rangle \varepsilon L . \equiv . \langle zxy \rangle \varepsilon L . \equiv . \langle xzy \rangle \varepsilon L$

(follows immediately from 9.62)

9.623 $Q_i` \bar{x} \varepsilon L$ for $i = 5, 6, \ldots, 8$

(follows from 9.62, 9.621)

9.63 $x \subseteq L . \supset . (\exists \bar{y})[x \subseteq \bar{y}]$

Proof: Consider $Od``x$, which is a set of ordinals; by 7.451 there exists an ordinal $\alpha$ greater than every element of $Od''x$. i.e., such that $Od``x \subseteq \alpha$. Moreover, such an α can be found with the additional restriction that $\alpha \varepsilon \mathfrak{W}(J_0)$, [e.g. by taking $J_0`\langle 0\alpha \rangle$ instead of $\alpha$ since $J_0` \langle 0\alpha \rangle \geq \alpha$ by 9.25] hence $F`\alpha = F``\alpha$, by 9.35, but $x \subseteq F``\alpha$, hence $x \subseteq F`\alpha$, and $F`\alpha$ is a constructible set. It follows that a constructible class which is a set is a constructible set, i.e.:

9.64 $\mathfrak{M}(\bar{X}) . \supset . \bar{X} \varepsilon L$.

Proof: By 9.41 and 9.63, $\bar{X}$ is contained in some $\bar{y}$. Therefore $\bar{X} \cdot \bar{y} = \bar{X}$, but $\bar{X} \cdot \bar{y}$ is a constructible set by 9.41.

9.65 $\mathfrak{L}(\bar{x})$

Proof: By 9.51, $\bar{x} \subseteq L$, by 9.61, $\bar{x} \cdot \bar{y} \varepsilon L$ for any $\bar{y}$.

9.66 $\bar{x} + \bar{y} \varepsilon L$

Proof: There is a $\bar{z}$ such that $\bar{x} + \bar{y} \subseteq \bar{z}$, by 9.51 and 9.63. $\bar{x} + \bar{y} = \bar{z} - [(\bar{z} - \bar{x}) - \bar{y}]$. Hence 9.6 gives the theorem.

9.8 $0 \varepsilon L$

Proof: $0 = \bar{x} - \bar{x}$, hence constructible, by 9.6.

9.81 $\mathfrak{L}(L)$

Proof: $L \subseteq L$, and because of 9.51, $\bar{x} \cdot L = \bar{x}$, hence $\bar{x} \cdot L \varepsilon L$. Therefore $\mathfrak{L}(L)$, by 9.41.

9.82 $\mathfrak{L}(E \cdot L)$

Proof: $E \cdot L \subseteq L$, also $\bar{x} \cdot E \varepsilon L$ by 9.6 since $X \cdot E$ is a fundamental operation, but $\bar{x} \cdot E = \bar{x} \cdot E \cdot L$ because $\bar{x} \subseteq L$, hence $\bar{x} \cdot E \cdot L \varepsilon L$, and so by 9.41, $\mathfrak{L}(E \cdot L)$.

9.83 $\mathfrak{L}(\bar{A} - \bar{B})$

Proof: $\bar{A} - \bar{B} \subseteq L$, moreover $\bar{x} \cdot \bar{A} - \bar{x} \cdot \bar{B}$ is constructible, by 9.41 and 9.6, but $\bar{x} \cdot \bar{A} - \bar{x} \cdot \bar{B} = \bar{x} \cdot (\bar{A} - \bar{B})$, hence $\bar{x} \cdot (\bar{A} - \bar{B}) \varepsilon L$, so that $\mathfrak{L}(\bar{A} - \bar{B})$, by 9.41.
Similarly:

9.84 $\mathfrak{L}(\bar{A} \cdot \bar{B})$,

and

9.85 $\mathfrak{L}(\bar{A} + \bar{B})$.
9.86 $\mathfrak{L}(Q_i``\bar{A})$; $i=5,\dots,8$

and

9.87 $\mathfrak{L}(L \cdot Q_4``\bar{A})$.

The last two theorems are proved as follows: $Q_5, \dots, Q_8$ take constructible sets into constructible sets, by 9.623, therefore

Q_i" $\bar{A} \subseteq L$, $i=5,\dots,8$. In order to prove that $\bar{x}\cdot Q_i$" $\bar{A} \varepsilon L$ for $i=4,\dots,8$, consider an arbitrary $y \varepsilon \bar{x}\cdot Q_i$" $\bar{A}$, $i=4,\dots,8$. y is an image by Q_i of some element of $\bar{A}$; take the element y' of $\bar{A}$ of lowest order of which y is an image. The totality of these y' for all elements y of $\bar{x}\cdot Q_i$" $\bar{A}$ is a set u of constructible sets and $u \subseteq \bar{A}$. By 9.63 we have $u \subseteq \bar{z}$, for some $\bar{z}$. $\bar{z}$ can be determined so that $\bar{z} \subseteq \bar{A}$, merely by taking $\bar{z}\cdot\bar{A}$. Hence we can assume: $u \subseteq \bar{z} \subseteq \bar{A}$. Therefore $\bar{x}\cdot Q_i$" $\bar{z} \subseteq \bar{x}\cdot Q_i$" $\bar{A}$ by 4.86, but also $\bar{x}\cdot Q_i$" $\bar{A} \subseteq \bar{x}\cdot Q_i$" $\bar{z}$ because any element of $\bar{x}\cdot Q_i$" $\bar{A}$ has an original in u hence in $\bar{z}$. Hence $\bar{x}\cdot Q_i$" $\bar{A}=\bar{x}\cdot Q_i$" $\bar{z}$, but $(\bar{x}\cdot Q_i$" $\bar{z})\varepsilon L$ by 9.6.

By means of theorems 4.92 to 4.96, theorems 9.86 and 9.87 take the following three forms:

9.871 $\mathfrak{L}[\mathfrak{D}(\bar{A})]$,
9.872 $\mathfrak{L}[\mathfrak{Cnv}_k(\bar{A})]$,
9.873 $\mathfrak{L}[L\cdot(\bar{V}\times\bar{A})]$.
9.88 $\mathfrak{L}(\bar{A}\times\bar{B})$

Proof: By 4.871 $\bar{A}\times\bar{B}=(V\times\bar{B})\cdot(\bar{A}\times V)=L\cdot(V\times\bar{B})\cdot L\cdot(\bar{A}\times V)$ because $\bar{A}\times\bar{B}\subseteq L$, by 9.62. By 9.873 and 9.872, $\mathfrak{L}[L\cdot(V\times\bar{B})]$ and $\mathfrak{L}[L\cdot(\bar{A}\times V)]$. Hence, by 9.84 $\mathfrak{L}(\bar{A}\times\bar{B})$.

9.89 $\mathfrak{L}[\mathfrak{W}(\bar{A})]$

Proof: $\mathfrak{W}(\bar{A})=\mathfrak{D}(\breve{\bar{A}})$, hence the result follows from 9.871 and 9.872.

9.90 $\mathfrak{L}[\bar{A}\restriction\bar{B}]$

Proof: $\bar{A}\restriction\bar{B}=\bar{A}\cdot(V\times\bar{B})=\bar{A}\cdot L\cdot(V\times\bar{B})$, hence the theorem, by 9.873 and 9.84.

9.91 $\mathfrak{L}[\bar{A}``\bar{B}]$

Proof: $\bar{A}``\bar{B}=\mathfrak{W}(\bar{A}\restriction\bar{B})$, hence the theorem, by 9.89 and 9.90.

9.92 $\mathfrak{L}(\{\overline{XY}\})$

Proof: By definition 3.1 $\{\overline{XY}\}$ is either 0 or $\{\bar{X}\}$ or $\{\bar{Y}\}$ or $\{\overline{XY}\}$ where now only sets can appear within the braces. Hence the theorem by 9.6, 9.65, 9.8.

Not all operations on constructible classes give necessarily constructible classes. For example, it cannot be shown that $\mathfrak{L}[\mathfrak{P}(\bar{X})]$.

Now consider <u>the model Δ</u> obtained as follows:

1. Class is construed as constructible class.
2. Set is construed as constructible set.
3. ε_1, the membership relation, is to be the ε-relation con-

confined to constructible classes, i.e., $\bar{X}\varepsilon_l\bar{Y}.\equiv.\bar{X}\varepsilon\bar{Y}$.

The operations, notions and special classes defined so far can be relativised for this model Δ by replacing in their definition or defining postulate the variables $X, Y, \ldots$, by $\bar{X}, \bar{Y}, \ldots$; the variables $x, y, \ldots$, by $\bar{x}, \bar{y}, \ldots$; ε by ε_l and the previously defined concepts and variables by the corresponding relativised ones, leaving the logical symbols [in particular also =, which is considered as a logical concept] as they stand. The relativised of a variable $\mathfrak{x}$ is a variable whose range is obtained by relativising the notion which defines the range of $\mathfrak{x}$. Note that for an operation or special class the relativised need not exist a priori, because the theorem which states existence and unicity (cf. p.12) may not hold in the model Δ, furthermore the relativised concept may depend on the particular definition which we chose, since equivalent definitions need not be equivalent in Δ. [However, as soon as we have proved that the axioms of Σ hold for Δ, we know that the relativised always does exist and does not depend on the particular definition.] If the relativised of a defined class A, operation $\mathfrak{A}$, notion $\mathfrak{B}$, variable $\mathfrak{x}$ exists (which presupposes that also the relativised of any symbol occurring in its definition exists), we denote it by $A_l, \mathfrak{A}_l, \mathfrak{B}_l, \mathfrak{x}_l$ [hence x_l , X_l have the same range as $\bar{x}$, $\bar{X}$]. $\mathfrak{A}_l$ and $\mathfrak{B}_l$ are defined for constructible classes as arguments only and we have the theorem:

10.1 If A_l and $\mathfrak{A}_l$ exist, then A_l is constructible and $\mathfrak{A}_l(\bar{X}_1, \ldots, \bar{X}_n)$ is constructible for any $\bar{X}_1, \ldots, \bar{X}_n$.

Evidently the relativised classes, notions, operations are at the same time classes, notions, operations of the system Σ , if the requirement on p.11 , that they be defined for any classes as argument, is met e.g. by stipulating that $\mathfrak{A}_l(X_1, \ldots, X_n)=0$ and $\mathfrak{B}_l(X_1, \ldots, X_n)$ is false, if $X_1, \ldots, X_n$ are not all constructible.

10 Dfn A special class A or operation $\mathfrak{A}$ or notion $\mathfrak{B}$ is called absolute, if A_l, $\mathfrak{A}_l$, or $\mathfrak{B}_l$, exists respectively and $A_l=A$, $\mathfrak{A}_l(\bar{X}_1, \ldots, \bar{X}_n)=\mathfrak{A}(\bar{X}_1, \ldots, \bar{X}_n)$, or $\mathfrak{B}_l(\bar{X}_1, \ldots, \bar{X}_n).\equiv.\mathfrak{B}(\bar{X}_1, \ldots, \bar{X}_n)$ respectively for any $\bar{X}_1, \ldots, \bar{X}_n$. A variable $\mathfrak{x}$ is called absolute, if the range of $\mathfrak{x}_l$ is the same as the range of $\mathfrak{x}$.

By theorem 10.1 we have:

10.11 If A (the operation $\mathfrak{A}$) is absolute then A is constructible ($\mathfrak{A}(\bar{X}_1, \ldots, \bar{X}_n)$ is constructible for any $\bar{X}_1, \ldots, \bar{X}_n$).

Concerning the meaning and purpose of the metamathematical notions of relativisation and absoluteness cf. introduction p.1 . The relativised of a propositional function φ or a proposition ψ

is denoted by φ_l , ψ_l , respectively and obtained by replacing any concept and variable occurring in it by the relativised one (presupposing that they all exist). In particular also the relativised of a theorem is quoted by putting a subscript l to its number.

10.12 ε is absolute.

This is true by definition of ε_l .

10.13 "$\subseteq$" is absolute.

Proof: $\bar{X} \subseteq_l \bar{Y} . \equiv . (\bar{u})[\bar{u}\varepsilon_l \bar{X} . \supset . \bar{u}\varepsilon_l \bar{Y}] . \equiv . (\bar{u})[\bar{u}\varepsilon\bar{X} . \supset . \bar{u}\varepsilon\bar{Y}]$
Also $\bar{X} \subseteq \bar{Y} . \equiv . (u)[u\varepsilon\bar{X} . \supset . u\varepsilon\bar{Y}]$. If $(u)[u\varepsilon\bar{X} . \supset . u\varepsilon\bar{Y}]$ then in particular $(\bar{u})[\bar{u}\varepsilon\bar{X} . \supset . \bar{u}\varepsilon\bar{Y}]$. On the other hand, the reverse implication holds, since, if u is not in L the condition holds vacuously, because the hypothesis $u\varepsilon\bar{X}$ is false. Therefore $\bar{X} \subseteq_l \bar{Y} . \equiv . \bar{X} \subseteq \bar{Y}$.

10.131 $\bar{X} \subseteq_l \bar{Y} . \bar{Y} \subseteq_l \bar{X} : \supset . \bar{X} = \bar{Y}$ i.e. the relativised axiom of extensionality holds.

Proof by 10.13 and the axiom of extensionality.

10.14 "$\subset$" is absolute.

Proof: $\bar{X} \subset_l \bar{Y} . \equiv : \bar{X} \subseteq_l \bar{Y} . \bar{X} \neq \bar{Y} . \equiv . \bar{X} \subset \bar{Y}$, by 10.13.
Similarly:

10.15 $\mathfrak{Cx}$ is absolute, i.e. $\mathfrak{Cx}_l(\bar{X}, \bar{Y}) . \equiv . \mathfrak{Cx}(\bar{X}, \bar{Y})$.
10.16 $\mathfrak{Em}$ is absolute, i.e. $\mathfrak{Em}_l(\bar{X}) . \equiv . \mathfrak{Em}(\bar{X})$.
10.17 The operation $\{\bar{X}\bar{Y}\}$ is absolute.

Proof: By 3.1 $\{\bar{X}\bar{Y}\}_l$ is the constructible class $\bar{Z}$ such that $(\bar{u})[\bar{u}\varepsilon\bar{X} . \equiv : \bar{u} = \bar{X} . \vee . \bar{u} = \bar{Y}]$. $\{\bar{X}\bar{Y}\}$ satisfies this condition on $\bar{Z}$ because it satisfies it even with (u) instead of $(\bar{u})$. Moreover, $\{\bar{X}\bar{Y}\}$ is constructible, by 9.92. Also $\{\bar{X}\bar{Y}\}$ is the only constructible class satisfying the condition (by 10.131). Hence the relativised of the operation $\{XY\}$ exists and $\{\bar{X}\bar{Y}\}_l = \{\bar{X}\bar{Y}\}$ for any $\bar{X}$, $\bar{Y}$, i.e. $\{XY\}$ is absolute.

10.18 If $\mathfrak{C}$ is defined by $\mathfrak{C}(\bar{X}) = \mathfrak{A}(\mathfrak{B}(\bar{X}))$ and $\mathfrak{A}$ and $\mathfrak{B}$ are absolute, then $\mathfrak{C}$ is absolute.

Proof: $\mathfrak{A}(\mathfrak{B}(\bar{X})) = \mathfrak{A}(\mathfrak{B}_l(\bar{X}))$, but $\mathfrak{B}_l(\bar{X})$ is constructible by 10.1, hence $\mathfrak{A}(\mathfrak{B}_l(\bar{X})) = \mathfrak{A}_l(\mathfrak{B}_l(\bar{X})) = \mathfrak{C}_l(\bar{X})$.
This principle holds also for operations with more than one argument.

10.19 The operation $\langle XY \rangle$ is absolute.

This is an immediate consequence of 10.17 and 10.18. Similarly:

10.20 The operation $\langle XYZ \rangle$ is absolute.
10.21 $\mathfrak{Un}$ is absolute.

Proof: $\mathfrak{Un}_l(\bar{X}) . \equiv . (\bar{u}\bar{v}\bar{w})[\langle\bar{v}\bar{u}\rangle_l \, \varepsilon_l \, \bar{X} . \langle\bar{w}\bar{u}\rangle_l \, \varepsilon_l \, \bar{X} : \supset . \bar{v}=\bar{w}]$.
By 10.12 and 10.19 the subscript l can be dropped wherever it appears on the right. The condition is now equivalent to that obtained by replacing $\bar{u}, \bar{v}, \bar{w}$, by u, v, w, respectively, as in the proof of 10.13 (using 9.62).

10.22 $\mathfrak{M}$ is absolute and $\mathfrak{Pr}$ is absolute.

Proof: $\mathfrak{M}_l(\bar{X}) . \equiv . \bar{X} \varepsilon L$, by definition of the model Δ on p.41 , therefore $\mathfrak{M}_l(\bar{X}) . \equiv . \mathfrak{M}(\bar{X})$, by 9.64 and axiom A2. Hence also: $\sim \mathfrak{M}_l(\bar{X}) . \equiv . \sim \mathfrak{M}(\bar{X})$.

Not all concepts can be proved to be absolute; for example, $\mathfrak{P}$ and V cannot be proved to be absolute.

10.23 $V_l = L$

Proof: V_l is defined by the postulate $(\bar{x})[\bar{x} \varepsilon V_l]$. L satisfies the condition, hence $L=V_l$, because of the relativised axiom of extensionality and because $\mathfrak{C}(L)$ by 9.81.

10.24 0 is absolute.

Proof: $(\bar{x})[\sim \bar{x} \varepsilon 0]$ and 0 is the only constructible class satisfying this postulate.

CHAPTER VI

PROOF OF THE AXIOMS OF GROUPS A-D FOR THE MODEL Δ

Every notion and operation appearing in the <u>axioms</u> has now been shown to be absolute. This facilitates the proofs of the relativised axioms, since in forming the relativised of a proposition all absolute notions and operations can be left as they stand, because by 10.1 only constructible classes can appear as their arguments, so that the relativised axioms may be formed merely by replacing X by $\bar{X}$ and x by $\bar{x}$. For convenience we list the axioms in their relativised form:

$A1_l$	$\mathfrak{L}(\bar{x})$,
2_l	$\bar{X}\varepsilon\bar{Y}.\supset.\mathfrak{M}(\bar{X})$,
3_l	$(\bar{u})[\bar{u}\varepsilon\bar{X}.\equiv.\bar{u}\varepsilon\bar{Y}].\supset.\bar{X}=\bar{Y}$,
4_l	$(\bar{x},\bar{y})(\exists\bar{z})(\bar{u})[\bar{u}\varepsilon\bar{z}.\equiv:\bar{u}=\bar{y}.\vee.\bar{u}=\bar{x}]$;
$B1_l$	$(\exists\bar{A})(\bar{x},\bar{y})[\langle\bar{x}\bar{y}\rangle\varepsilon\bar{A}.\equiv.\bar{x}\varepsilon\bar{A}]$,
2_l	$(\bar{A},\bar{B})(\exists\bar{C})(\bar{x})[\bar{x}\varepsilon\bar{C}.\equiv:\bar{x}\varepsilon\bar{A}.x\varepsilon\bar{B}]$,
3_l	$(\bar{A})(\exists\bar{B})(\bar{x})[\bar{x}\varepsilon\bar{B}.\equiv.\sim\bar{x}\varepsilon\bar{y}]$,
4_l	$(\bar{A})(\exists\bar{B})(\bar{x})[\bar{x}\varepsilon\bar{B}.\equiv.(\exists\bar{y})[\langle\bar{y}\bar{x}\rangle\varepsilon\bar{A}]]$,
5_l	$(\bar{A})(\exists\bar{B})(\bar{x},\bar{y})[\langle\bar{y}\bar{x}\rangle\varepsilon\bar{B}.\equiv.\bar{x}\varepsilon\bar{A}]$,
6_l	$(\bar{A})(\exists\bar{B})(\bar{x},\bar{y})[\langle\bar{x}\bar{y}\rangle\varepsilon\bar{B}.\equiv.\langle\bar{y}\bar{x}\rangle\varepsilon\bar{A}]$,
7_l	$(\bar{A})(\exists\bar{B})(\bar{x},\bar{y},\bar{z})[\langle\bar{x}\bar{y}\bar{z}\rangle\varepsilon\bar{B}.\equiv.\langle\bar{y}\bar{z}\bar{x}\rangle\varepsilon\bar{A}]$,
8_l	$(\bar{A})(\exists\bar{B})(\bar{x},\bar{y},\bar{z})[\langle\bar{x}\bar{y}\bar{z}\rangle\varepsilon\bar{B}.\equiv.\langle\bar{x}\bar{z}\bar{y}\rangle\varepsilon\bar{A}]$;
$C1_l$	$(\exists\bar{a})\{\sim\mathfrak{Em}(\bar{a}).(\bar{x})[\bar{x}\varepsilon\bar{a}.\supset.(\exists\bar{y})(\bar{y}\varepsilon\bar{a}.\bar{x}\subset\bar{y})]$,
2_l	$(\bar{x})(\exists\bar{y})(\bar{u},\bar{v})[\bar{u}\varepsilon\bar{v}.\bar{v}\varepsilon\bar{x}:\supset.\bar{u}\varepsilon\bar{y}]$,
3_l	$(\bar{x})(\exists\bar{y})(\bar{u})[\bar{u}\subseteq\bar{x}.\supset.\bar{u}\varepsilon\bar{y}]$,
4_l	$(\bar{x},\bar{A})\{\mathfrak{Un}(\bar{A}).\supset.(\exists\bar{y})(\bar{u})[\bar{u}\varepsilon\bar{y}.\equiv.(\exists\bar{v})(\bar{v}\varepsilon\bar{x}.\langle\bar{u}\bar{v}\rangle\varepsilon\bar{A})]$;
D_l	$\sim\mathfrak{Em}(\bar{A}).\supset.(\exists\bar{x})[\bar{x}\varepsilon\bar{A}.\mathfrak{Ex}(\bar{x},\bar{A})]$.

$A1_l$ is theorem 9.65, $A2_l$ is immediate from A2, $A3_l$ holds by 10.131, $A4_l$ is satisfied for $\bar{z}=\{\bar{x}\bar{y}\}$, which is constructible by 9.6. Now we prove $B1\text{-}8_l$ by exhibiting in each case a constructible class satisfying the conditions, as follows:

$B1_l$ Take $\bar{A}=E\cdot L$, the class $E\cdot L$ is constructible by 9.82 and satisfies $\langle\bar{x}\bar{y}\rangle\varepsilon E\cdot L.\equiv.\bar{x}\varepsilon\bar{y}$, because $\langle\bar{x}\bar{y}\rangle\varepsilon E.\equiv.\bar{x}\varepsilon\bar{y}$ and $\langle\bar{x}\bar{y}\rangle\varepsilon L$.

2 Take $\bar{C}=\bar{A}\cdot\bar{B}$, this class is constructible by 9.84 and satisfies B2.

3 Take $\bar{B}=L\text{-}\bar{A}$, this class is constructible by 9.83, 9.81 and satisfies B3.

4 Take $\bar{B}=\mathfrak{D}(\bar{A})$. By 9.871 $\mathfrak{D}(\bar{A})$ is constructible. $x\varepsilon\bar{B}.\equiv.(\exists y)[\langle yx\rangle\varepsilon\bar{A}]$. Therefore, in particular,

$\bar{x}\varepsilon\bar{B}.\equiv.(\exists y)[\langle y\bar{x}\rangle\varepsilon A].\equiv.(\exists\bar{y})[\langle\bar{y}\bar{x}\rangle\varepsilon\bar{A}]$. The last equivalence holds, because, if there exists a y it must be constructible by 9.62.

5_ι Take $\bar{B}=L\cdot(V\times A)$. $\bar{B}$ is constructible, by 9.873 . $\langle xy\rangle\varepsilon\bar{B}.\equiv.\langle xy\rangle\varepsilon L.y\varepsilon\bar{A}$ therefore $\langle\bar{x}\bar{y}\rangle\varepsilon\bar{B}.\equiv.\langle\bar{x}\bar{y}\rangle\varepsilon L.\bar{y}\varepsilon\bar{A}$ so that $\langle\bar{x}\bar{y}\rangle\varepsilon\bar{B}.\equiv.\bar{y}\varepsilon\bar{A}$, since $\langle\bar{x}\bar{y}\rangle\varepsilon L$, by 9.62.

6 Take $\bar{B}=\mathfrak{Cnv}(\bar{A})$. $\bar{B}$ is constructible, by 9.872. $\langle xy\rangle\varepsilon\mathfrak{Cnv}(A).\equiv.\langle yx\rangle\varepsilon\bar{A}$; therefore, in particular, $\langle xy\rangle\varepsilon\mathfrak{Cnv}(\bar{A}).\equiv.\langle\bar{y}\bar{x}\rangle\varepsilon\bar{A}$.

Axioms B7-8_ι are proved in the same manner. Now consider axioms C1-4_ι:

1_ι $C1_\iota$ is satisfied by $\bar{a}=F'\omega$.
Proof: $\omega\varepsilon\mathfrak{W}(J_0)$ by 9.27, hence $F'\omega=F''\omega$. If $\bar{x}\varepsilon\bar{a}$ (i.e. $\bar{x}=F'\alpha$, $\alpha<\omega$), let β be an integer $\varepsilon\mathfrak{W}(J_0)$ and $>\alpha$ (e.g. $\beta=J_0'\langle 0,\alpha+1\rangle$ by 9.25 and 9.26) and put $\bar{y}=F'\beta$, then $\bar{y}\varepsilon\bar{a}$ and $\bar{y}\supset\bar{x}$ because $F'\beta=F''\beta$ and $F'\alpha\subseteq F''\beta$. Moreover: $F'\alpha\varepsilon F''\beta$ but $\sim(F'\alpha\varepsilon F'\alpha)$ so that $F'\alpha\subset F'\beta$.

2_ι Consider $\mathfrak{S}(\bar{x})$; this is a set of constructible sets by 5.122 and 9.51. Therefore, by 9.63, there is a $\bar{y}$ such that $\mathfrak{S}(\bar{x})\subseteq\bar{y}$. Hence $(u,v)[u\varepsilon v.v\varepsilon\bar{x}:\supset.u\varepsilon\bar{y}]$, therefore $(\bar{u},\bar{v})[\bar{u}\varepsilon\bar{v}.\bar{v}\varepsilon\bar{x}:\supset.\bar{u}\varepsilon\bar{y}]$, that is, $\bar{y}$ satisfies the condition of C2.

3_ι Consider $L\cdot\mathfrak{P}(\bar{x})$ (which is a set by 5.121) and take $\bar{y}$ such that $L\cdot\mathfrak{P}(\bar{x})\subseteq\bar{y}$, by 9.63. Then $u\varepsilon L.\mathfrak{P}(\bar{x}).\supset.u\varepsilon\bar{y}$. Therefore $\bar{u}\varepsilon L\cdot\mathfrak{P}(\bar{x}).\supset.\bar{u}\varepsilon\bar{y}$, so that $\bar{u}\varepsilon\mathfrak{P}(\bar{x}).\supset.\bar{u}\varepsilon\bar{y}$, that is, $\bar{u}\subseteq\bar{x}.\supset.\bar{u}\varepsilon\bar{y}$.

4_ι Take $\bar{y}=\bar{A}''\bar{x}$. $\bar{y}$ is constructible, by 9.91. $u\varepsilon\bar{y}.\equiv.(\exists v)[v\varepsilon\bar{x}.\langle uv\rangle\varepsilon\bar{A}]$, therefore, in particular, $\bar{u}\varepsilon\bar{y}.\equiv.(\exists v)[v\varepsilon\bar{x}.\langle\bar{u}v\rangle\varepsilon\bar{A}]$. Now if there is a constructible v there is a v satisfying the condition; on the other hand, if there is a v , v will be constructible, since $v\varepsilon\bar{x}$. Therefore $\bar{u}\varepsilon\bar{y}.\equiv.(\exists\bar{v})[\bar{v}\varepsilon\bar{x}.\langle\bar{u}\bar{v}\rangle\varepsilon A]$.

Finally, consider axiom D_ι. By axiom D, $(\exists x)[x\varepsilon\bar{A}.\mathfrak{Ex}(x\bar{A})]$. But x is constructible, since $x\varepsilon\bar{A}$. Hence there is an $\bar{x}$ satisfying the condition.

Since all axioms of Σ hold in Δ, it follows now that all theorems proved so far also hold in the model Δ, except perhaps those based on the axiom of choice. Therefore the existence and unicity theorems necessary for the definition of the special classes and the operations introduced so far also will hold in Δ, and, as a result, the relativised of every concept introduced so far exists [except those definitions marked by * which depend on the axiom of choice]; in particular also $\mathfrak{L}_\iota$ and L_ι exist.

CHAPTER VII

PROOF THAT V = L HOLDS IN THE MODEL Δ

In order to prove that the axiom of choice and the generalised continuum-hypothesis hold for the model Δ, we shall show: 1.) that both of them follow from the axioms of Σ and the additional axiom $V=L$ (which says that every set is constructible) and 2.) that $V=L$ holds in the model Δ, i.e. $V_l = L_l$. We begin with item 2.). Since $V_l = L$ by 10.23, it is sufficient to prove $L_l = L$, that is, <u>the class of constructible sets is absolute</u>. To that end, it will be shown that all operations, etc. used in the construction of L are absolute.

A general remark for proofs of absoluteness will be useful. In order for the operation $\mathfrak{A}(X_1,\dots,X_n)$ to be absolute it is sufficient to show that

(1) $\mathfrak{A}$ gives constructible classes when applies to constructible classes, and

(2) $\mathfrak{A}$ satisfies the relativised defining postulate i.e. if $\mathfrak{A}$ is defined by $\varphi(\mathfrak{A}(X_1,\dots,X_n),X_1,\dots,X_n)$,
then $\varphi_l(\mathfrak{A}(\bar{X}_1,\dots,\bar{X}_n),\bar{X}_1,\dots,\bar{X}_n)$.

It is easily verified that (1) and (2) are sufficient, namely, as follows: $\mathfrak{A}_l$ exists, since the model satisfies the axioms of Σ. Hence φ_l has the property that for any $\bar{X}_1,\dots,\bar{X}_n$ there exists at most one $\bar{Y}$ such that $\varphi_l(\bar{Y},\bar{X}_1,\dots,\bar{X}_n)$. But $\varphi_l(\mathfrak{A}_l(\bar{X}_1,\dots,\bar{X}_n),\bar{X}_1,\dots,\bar{X}_n)$ by definition of $\mathfrak{A}_l$ and $\varphi_l(\mathfrak{A}(\bar{X}_1,\dots,\bar{X}_n),\bar{X}_1,\dots,\bar{X}_n)$ by assumption (2). Therefore $\mathfrak{A}_l(\bar{X}_1,\dots,\bar{X}_n)=\mathfrak{A}(\bar{X}_1,\dots,\bar{X}_n)$. Similarly for the particular class A it is sufficient to show that it is constructible and satisfies the relativised postulate. Remember also that by 10.18 operations defined by substituting absolute operations into absolute operations are absolute.

11.1 "$\times$" is absolute.

Proof: $\bar{A}\times\bar{B}$ is constructible, by 9.88.
$u \varepsilon \bar{A}\times\bar{B} . \equiv . (\exists v,w)[v\varepsilon\bar{A} . w\varepsilon\bar{B} . u=\langle vw\rangle]$ by definition 4.1. Therefore $\bar{u} \varepsilon \bar{A}\times\bar{B} . \equiv . (\exists v,w)[v\varepsilon\bar{A} . w\varepsilon\bar{B} . \bar{u}=\langle vw\rangle]$. Now, in the usual manner, the condition on the right is equivalent to that obtained by replacing v,w by $\bar{v},\bar{w}$ respectively. Therefore $\bar{A}\times\bar{B}$ satisfies the relativised postulate, hence "$\times$" is absolute, by the remark made above.

11.11 The operations A^2, A^3, ... , are absolute.

This follows from 10.18 and 11.1.

11.12 $\mathfrak{Rel}$ and $\mathfrak{Rel}_3$ are absolute.

Proof: $\mathfrak{Rel}(\bar{X}) . \equiv . \bar{X} \subseteq V^2$ and $\mathfrak{Rel}_l(\bar{X}) . \equiv . \bar{X} \subseteq L^2$ by 10.23 but $\bar{X} \subseteq L^2 . \equiv . \bar{X} \subseteq V^2$, by 9.62. Hence $\mathfrak{Rel}_l(\bar{X}) . \equiv . \mathfrak{Rel}(\bar{X})$. Similarly for $\mathfrak{Rel}_3$.

11.13 $\mathfrak{D}$ is absolute.

Proof: $\mathfrak{D}(\bar{A})$ is constructible, by 9.871. $x \varepsilon \mathfrak{D}(A) . \equiv . (\exists y)[\langle yx \rangle \varepsilon \bar{A}]$, therefore $\bar{x} \varepsilon \mathfrak{D}(\bar{A}) . \equiv . (\exists y)[\langle y\bar{x} \rangle \varepsilon \bar{A}]$. In the usual way, the last condition is equivalent to that obtained by replacing y by $\bar{y}$, so that $\mathfrak{D}(A)$ satisfies the relativised postulate.

11.14 "$\cdot$" is absolute.

Proof: $\bar{A} \cdot \bar{B}$ is constructible, by 9.84. $x \varepsilon \bar{A} \cdot \bar{B} . \equiv : x \varepsilon \bar{A} . x \varepsilon \bar{B}$, therefore $\bar{x} \varepsilon \bar{A} \cdot \bar{B} . \equiv : \bar{x} \varepsilon \bar{A} . \bar{x} \varepsilon \bar{B}$, that is, $\bar{A} \cdot \bar{B}$ satisfies the relativised postulate.

11.15 $\mathfrak{Cnv}_k$ is absolute $(k=1,2,3)$.

Proof: $\mathfrak{Cnv}_k(\bar{A})$ is constructible, by 9.872. Consider e.g. $\mathfrak{Cnv}_1(\bar{A})$. It satisfies the condition

$$\mathfrak{Rel}\,(\mathfrak{Cnv}_1(\bar{A})) . (x,y)[\langle xy \rangle \varepsilon\, \mathfrak{Cnv}_1(\bar{A}) . \equiv . \langle yx \rangle \varepsilon \bar{A}]$$

by definition. This condition implies the relativised statement by 11.12. Similarly for $\mathfrak{Cnv}_k(\bar{A})$.

11.16 "$\upharpoonright$" is absolute.

Proof: $\bar{A} \upharpoonright \bar{B} = \bar{A} \cdot (V \times \bar{B})$ and $\bar{A} \upharpoonright_l \bar{B} = \bar{A} \cdot (L \times \bar{B})$. But $\bar{A} \cdot (V \times \bar{B}) \subseteq L \times L$ by 9.62, therefore $\bar{A} \upharpoonright \bar{B} = \bar{A} \cdot (V \times \bar{B}) \cdot (L \times L) = \bar{A} \cdot (L \times \bar{B})$, by 4.87. Therefore $\bar{A} \upharpoonright \bar{B} = \bar{A} \upharpoonright_l \bar{B}$.

11.17 "$\mathfrak{W}$" is absolute.

Proof: $\mathfrak{W}(A) = \mathfrak{D}(\mathfrak{Cnv}(A))$ by definition. Hence the theorem by 10.18, 11.13, and 11.15.

11.18 The operation $A``B$ is absolute.

Proof: $A``B = \mathfrak{W}(A \upharpoonright B)$, by definition. Hence the theorem by 10.18.

11.181 The relativised operation of the complement is $L - \bar{X}$.

Proof: $L - \bar{X}$ is constructible by 9.81, 9.83 and $\bar{y} \varepsilon L - \bar{X} . \equiv . \sim \bar{y} \varepsilon \bar{X}$.

11.19 The operation $A - B$ is absolute.

Proof: $\bar{A}-_l\bar{B}=\bar{A}\cdot(L-\bar{B})=\bar{A}\cdot L\cdot(-\bar{B})=\bar{A}\cdot(-\bar{B})=\bar{A}-\bar{B}$.

11.20 "+" is absolute.

Proof: $\bar{A}+_l\bar{B}=L-[(L-\bar{A})\cdot(L-\bar{B})]=L-[L-(\bar{A}+\bar{B})]=\bar{A}+\bar{B}$ (since $\bar{A}+\bar{B}\subseteq L$).

11.21 $E_l=E\cdot L$

Proof: E·L is constructible by 9.82. Also

$$\mathfrak{Rel}_l(E\cdot L).(\bar{x},\bar{y})[\langle\bar{x}\bar{y}\rangle\varepsilon E\cdot L.\equiv.\bar{x}\varepsilon\bar{y}] \quad ,$$

since $\mathfrak{Rel}(E\cdot L)$ and since $\langle\bar{x}\bar{y}\rangle\,\varepsilon\, L$ and $\langle\bar{x}\bar{y}\rangle\varepsilon E.\equiv.\bar{x}\varepsilon\bar{y}$. Therefore E·L satisfies the relativised postulate.

11.22 $\mathfrak{F}_2$ is absolute.

Proof: $\mathfrak{F}_{2_l}(\bar{X},\bar{Y})=\bar{X}\cdot_l E_l=\bar{X}\cdot L\cdot E=\bar{X}\cdot E=\mathfrak{F}_2(\bar{X},\bar{Y})$, by 11.14, 11.21.

11.221 All the fundamental operations $\mathfrak{F}_i$ $(i=1,2,\ldots,8)$ are absolute.

The proof follows from 10.17, 11.22, 11.19, 11.16, 11.13, 11.15 respectively, using 10.18 and 11.14.

11.23 The binary operation $A\text{‘}X$ is absolute.

Proof: Since any y satisfying $\langle y\bar{X}\rangle\varepsilon\bar{A}$ is constructible by 9.51, we have: if there is exactly one <u>constructible</u> set y such that $\langle y\bar{X}\rangle\varepsilon\bar{A}$, there is exactly one set, and vice versa. Therefore $\bar{A}\text{‘}_l\,\bar{X}=\bar{A}\text{‘}\bar{X}$ in this case; in the contrary case both are 0.

11.3 $\mathfrak{Comp}$ is absolute.

Proof:

$$\begin{aligned}\mathfrak{Comp}(X).&\equiv.(u)[u\varepsilon\bar{X}.\supset u\subseteq\bar{X}].\\ &\equiv.(\bar{u})[\bar{u}\varepsilon\bar{X}.\supset\bar{u}\subseteq\bar{X}].\equiv.\mathfrak{Comp}_l(X)\ .\end{aligned}$$

11.31 $\mathfrak{Ord}$ is absolute.

Proof:

$$\begin{aligned}\mathfrak{Ord}(\bar{X}).&\equiv:\mathfrak{Comp}(\bar{X}).(u,v)[u,v\varepsilon\bar{X}:\supset:.u=v.\vee.u\varepsilon v.\vee.v\varepsilon u].\\ &\equiv:\mathfrak{Comp}_l(\bar{X}).(\bar{u},\bar{v})[\bar{u}.\bar{v}\varepsilon\bar{X}:\supset:.\bar{u}=\bar{v}.\vee.\bar{u}\varepsilon\bar{v}.\vee.\bar{v}\varepsilon\bar{u}].\\ &\equiv:\mathfrak{Ord}_l(\bar{X})\ .\end{aligned}$$

The first and last equivalences follow immediately from the definition of $\mathfrak{Ord}$ and $\mathfrak{Ord}_l$.

11.32 $\mathfrak{O}$ is absolute.

Proof:

$$\begin{aligned}\mathfrak{O}(\bar{X}).&\equiv:\mathfrak{Ord}(\bar{X}).\mathfrak{M}(\bar{X}):\equiv:\mathfrak{Ord}_l(\bar{X}).\mathfrak{M}_l(\bar{X}).\\ &\equiv:\mathfrak{O}_l(\bar{X})\ ,\text{ by 11.31, 10.22.}\end{aligned}$$

11.31 says that the ordinals of the model Δ are the same as the ordinals which belong to the model Δ. This does not mean that the ordinals of the model are the same as the ordinals of the original system, since nothing is said of those ordinals which may not belong to the model (i.e. may not be constructible). Cf. however 11.42.

11.4 "$\mathfrak{Fnc}$" is absolute.

Proof: $\mathfrak{Fnc}_l(\bar{Y}) . \equiv : \mathfrak{Rel}_l(\bar{Y}) . \mathfrak{Un}_l(\bar{Y}) .$
$\equiv : \mathfrak{Rel}(\bar{Y}) . \mathfrak{Un}(\bar{Y})$ by 11.12, 10.21, 4.61.

11.41 "$\mathfrak{Fn}$" is absolute.

Proof: $\bar{Y} \mathfrak{Fn}_l \bar{X} . \equiv : \mathfrak{Fnc}_l(\bar{Y}) . \mathfrak{D}_l(\bar{Y}) = \bar{X} .$
$\equiv : \mathfrak{Fnc}(\bar{Y}) . \mathfrak{D}(\bar{Y}) = \bar{X}$ by 11.4, 11.13, 4.63.

11.42 On is absolute.

Proof: $\mathfrak{Ord}_l(On_l)$ by 7.16_l and $\mathfrak{Pr}_l(On_l)$ by 7.17_l. But On_l is constructible by 10.1, hence $\mathfrak{Ord}(On_l)$ and $\mathfrak{Pr}(On_l)$ because $\mathfrak{Ord}$ and $\mathfrak{Pr}$ are absolute by 11.31 and 10.22. Hence $On_l = On$ by 7.2.

By 10.11 it follows from 11.42 that $On \subseteq L$, in other words, every ordinal number is constructible. Furthermore 11.42 implies:

11.421 The variables $\alpha, \beta, \ldots,$ are absolute.
11.43 "$<$" is absolute.

Proof: "$<$" is by definition the same as "ε".

11.44 "$\leq$" is absolute.

Proof: $X \leq Y$ is by definition $X \varepsilon Y . \vee . X = Y$.

11.45 "$\dot{+}1$" is absolute.

Proof: 7.4, 10.17, 11.20 and 10.18.

11.451 Each of the symbols 0,1,2,3,..., etc. is absolute.

Proof by 10.24 and 11.45.

11.46 $\mathfrak{S}$ (and therefore $\mathfrak{Max}$ and $\mathfrak{Lim}$) is absolute.

Proof: $z \varepsilon \mathfrak{S}(\bar{X}) . \equiv . (\exists v)[z \varepsilon v . v \varepsilon \bar{X}] . \equiv . (\exists \bar{v})[z \varepsilon \bar{v} . \bar{v} \varepsilon \bar{X}] . \equiv . z \varepsilon \mathfrak{S}_l(\bar{X})$. Therefore $\mathfrak{S}(\bar{X}) = \mathfrak{S}_l(\bar{X})$ by the axiom of extensionality.

What is left now is to show that the special classes R, S, J, K_1, K_2, F, and finally L, are absolute, where R is the ordering for pairs defined in 7.81, S is the ordering of the triples $\langle i\alpha\beta\rangle$ defined by 9.2, and F is the function introduced by 9.3 which defines L. For each of these the proof of absoluteness will be based on the following lemma:

If the class A is defined by the postulate $\varphi(A)$ and if all defined classes, operations, notions, and variables appearing in φ are absolute, then A is absolute.

Proof: If φ satisfies the condition above then $\varphi_l(\bar{X}) . \equiv . \varphi(\bar{X})$. Also $\varphi_l(A_l)$ and $\varphi(A)$ by definition of A_l and A. Since by 10.1 A_l is constructible, $\varphi_l(A_l)$ implies $\varphi(A_l)$, hence $A_l = A$, because both $\varphi(A_l)$ and $\varphi(A)$.

11.5 "R" is absolute.

Proof: By definition 7.81 we have
$R \subseteq (On^2)^2 . (\alpha,\beta,\gamma,\delta)[\langle\langle\alpha\beta\rangle\langle\gamma\delta\rangle\rangle \varepsilon R .$
$\equiv :: . \mathfrak{Max}\{\alpha\beta\} < \mathfrak{Max}\{\gamma\delta\} . \vee :: \mathfrak{Max}\{\alpha\beta\} = \mathfrak{Max}\{\gamma\delta\} : . \beta < \delta . \vee : \beta = \delta . \alpha < \gamma]$.
The following concepts appear in the defining postulate: $\subseteq$, On, ^{2}, $\langle\ \rangle$, $\mathfrak{Max}$, $\{\ \}$, $<$, ε , and variables α, β, ... , all of which have been proved absolute by 10.13, 11.42, 11.11, 10.19, 11.46, 10.17, 11.43, 10.12, 11.421 respectively.

11.51 "S" is absolute.

Proof: By definition 9.2 we have
$S \subseteq (9 \times On^2)^2 . (\alpha,\beta,\gamma,\delta,\mu,\nu)\{\mu < 9 . \nu < 9 : \supset$
$:: \langle\langle\mu\alpha\beta\rangle\langle\nu\gamma\delta\rangle\rangle \varepsilon S . \equiv : . \langle\langle\alpha\beta\rangle\langle\gamma\delta\rangle\rangle \varepsilon R . \vee : \langle\alpha\beta\rangle = \langle\gamma\delta\rangle . \mu < \nu\}$.
In the postulate for S the following concepts appear, other than those appearing previously in the postulate for R: $\times$, R, 9, which are absolute by 11.1, 11.5, 11.451, respectively.

11.52 "J" is absolute.

Proof: By definition 9.21 we have
$J \,\mathfrak{Fn}(9 \times On^2) . \mathfrak{W}(J) = On . (\alpha,\beta,\gamma,\delta,\mu,\nu)[\mu,\nu < 9 . \supset$
$: \langle\mu\alpha\beta\rangle S \langle\nu\gamma\delta\rangle . \supset . J`\langle\mu\alpha\beta\rangle < J`\langle\nu\gamma\delta\rangle]$.
The only additional symbols in this postulate are: $\mathfrak{Fn}$, $\mathfrak{W}$, and ‘ , all of which have been proved absolute by 11.41, 11.17, 11.23 respectively.

11.53 Each "J_i" is absolute; i=0,1,2,...,8 .

Proof: $J_0`\langle\alpha\beta\rangle = J`\langle 0\alpha\beta\rangle$. $J_0 \,\mathfrak{Fn}\, On^2$. Here there are no symbols but those mentioned before. Similarly for $J_1,\ldots,J_8$.

11.54 K_1 and K_2 are absolute.

Proof: In the defining postulate 9.24 there are no symbols but those mentioned before.

11.6 "F" is absolute.

Proof: The only additional symbols appearing in the defining postulate 9.3 are $\upharpoonright$ and $\mathfrak{F}_1, \ldots, \mathfrak{F}_8$ which are absolute by 11.16, 11.221 respectively.

11.7 "L" is absolute.

Proof: $L = \mathfrak{W}(F)$ and $\mathfrak{W}$, F are absolute by 11.6, 11.17.

It has now been demonstrated from the axioms of Σ that $L_\iota = L$, hence also that $V_\iota = L_\iota$ i.e. that the proposition V=L holds in the model Δ. This proves that if there exists a model for the axioms of groups A, B, C, D there exists also a model for the augmented set of axioms obtained by adding as an axiom the proposition V=L, namely the model consisting of the classes and sets "constructible" in the given model for Σ. Thus if the system A, B, C, D is consistent, the augmented system is consistent. Another way of putting this argument is as follows: If a contradiction were obtained from V=L and the axioms of Σ, (i.e. the axioms of groups A, B, C, D) then the same contradiction could be derived also from $V_\iota = L_\iota$ and the relativised axioms A_ι, B_ι, C_ι, D_ι. But $V_\iota = L_\iota$ and A_ι, B_ι, C_ι, D_ι can be proved in Σ as shown before, hence Σ would be contradictory, and a contradiction in Σ could actually be constructed if a contradiction from Σ and V=L were given.

CHAPTER VIII

PROOF THAT V = L IMPLIES THE AXIOM OF CHOICE AND THE GENERALISED CONTINUUM-HYPOTHESIS

Now it remains only to be shown that the axiom of choice and the generalised continuum-hypothesis follow from $V=L$ and Σ.

For the axiom of choice this is immediate since the relation As defined in 11.8, which singles out the element of least order in any non-vacuous constructible set, evidently satisfies axiom E if $V=L$.

11.8 Dfn $\langle yx \rangle \varepsilon As . \equiv : y \varepsilon x . (z)[Od`z < Od`y . \supset . \sim z \varepsilon x] . \mathfrak{Rel}(As)$

$As`x$ is what may be called the "designated" element of x.

11.81 Dfn $C`\alpha = Od`[As`(F`\alpha)] . C \mathfrak{Fn} On$

$C`\alpha$ is the order of the "designated" element of $F`\alpha$. Hence $C`\alpha \leq \alpha$ by 9.52.

The rest of these lectures is devoted to the derivation of the generalised continuum-hypothesis from $V=L$ and the axioms of Σ. Since we have just derived the axiom of choice from these assumptions, we are justified in using all starred theorems and definitions in this derivation. The theorems which follow from now on are only claimed to be consequences of Σ and $V=L$. However only 12.2 really depends on $V=L$, in all the others $V=L$ is not used and even the axiom of choice could be avoided in their proofs, if one wanted to.

12.1 $\overline{\overline{F``\omega_\alpha}} = \omega_\alpha$

Proof: $\overline{\overline{F``\omega_\alpha}} \leq \overline{\overline{\omega_\alpha}} = \omega_\alpha$ by 8.31. On the other hand, there exists a subset of ω_α, namely $\omega_\alpha \cdot \mathfrak{W}(J_0)$, such that the values of F over this subset are all different, since if $\gamma \neq \delta$, and $\gamma, \delta \varepsilon \omega_\alpha \cdot \mathfrak{W}(J_0)$, assume $\gamma < \delta$, then $F`\gamma \varepsilon F`\delta$, by 9.3, so that $F`\gamma \neq F`\delta$. But $\overline{\overline{\omega_\alpha \cdot \mathfrak{W}(J_0)}} \geq \omega_\alpha$, because $\overline{\overline{J_0``(\omega_\alpha{}^2)}} \subseteq \omega_\alpha \cdot \mathfrak{W}(J_0)$ by 9.26 and $J_0$ is one to one. Hence $\overline{\overline{F``\omega_\alpha}} \geq \omega_\alpha$.

By 12.1 the generalised continuum-hypothesis follows immediately from the following theorem:

12.2 $\mathfrak{P}(F``\omega_\alpha) \subseteq F``\omega_{\alpha+1}$.

This theorem is proved by means of the following lemma:

12.3 If $m \subseteq On$ and m is closed with respect to C, K_1, K_2 and with respect to $J_0,\ldots,J_8$ as triadic relations and if G is an isomorphism from m to an ordinal o with respect to E, then G is also an isomorphism with respect to $\hat\alpha\hat\beta[F`\alpha\,\varepsilon\,F`\beta]$ i.e. $\alpha,\beta\,\varepsilon\,m \supset [F`\alpha\,\varepsilon\,F`\beta . \equiv . F`G`\alpha\,\varepsilon\,F`G`\beta]$.

We show first that 12.3 implies 12.2.

Proof: Consider $u\,\varepsilon\,\mathfrak{P}(F``\omega_\alpha)$ , that is $u \subseteq F``\omega_\alpha$. By V=L there is a δ such that $u=F`\delta$; form the closure of the set $\omega_\alpha + \{\delta\}$, with respect to C, K_1, K_2, and with respect to J_i, i=0,1,...,8, as triadic relations, according to 8.73, and let the closure be denoted by m . Now by 8.73, m is a set and $\bar{\bar{m}} = \omega_\alpha$. m is a set of ordinals, hence m is well-ordered by E by 7.161 and is isomorphic to some ordinal number o by 7.7. Let the isomorphism be denoted by G, so that $G``m=o$. For brevity, let α' denote $G`\alpha$. By lemma 12.3 we have:

$$\alpha,\beta\,\varepsilon\,m . \supset : F`\alpha\,\varepsilon\,F`\beta . \equiv . F`\alpha'\,\varepsilon\,F`\beta' \; .$$

Now consider δ', the image of δ by G. $\delta'\,\varepsilon\,o$, that is, $\delta' < o$. Since G is one to one as an isomorphism, $\bar{\bar{o}} = \bar{\bar{m}} = \omega_\alpha$, from which it follows that $o < \omega_{\alpha+1}$, hence $\delta' < \omega_{\alpha+1}$. Also, for any $\beta\,\varepsilon\,m$, $F`\beta\,\varepsilon\,F`\delta . \equiv . F`\beta'\,\varepsilon\,F`\delta' . \omega_\alpha \subseteq m$, by definition and ω_α is complete (as an ordinal number). Therefore ω_α is an E-section of m , hence ω_α is mapped by G on an E-section of o i.e. by 7.21 on an ordinal number. But by 7.62, this can be only the identical mapping of ω_α onto itself. Therefore, if $\beta\,\varepsilon\,\omega_\alpha$, then $\beta' = \beta$. Hence $F`\beta\,\varepsilon\,F`\delta . \equiv . F`\beta\,\varepsilon\,F`\delta'$, for $\beta\,\varepsilon\,\omega_\alpha$ that is, $F`\delta$ and $F`\delta'$ have exactly the same elements with $F``\omega_\alpha$ in common, i.e., $F`\delta \cdot F``\omega_\alpha = F`\delta' \cdot F`\omega_\alpha$ but $F`\delta \subseteq F``\omega_\alpha$, by assumption therefore $F`\delta = F`\delta' \cdot F``\omega_\alpha$. But $\omega_\alpha\,\varepsilon\,\mathfrak{W}(J_0)$ by 9.27, therefore, by 9.35, $F``\omega_\alpha = F`\omega_\alpha$ , hence $u = F`\delta = F`\delta' \cdot F`\omega_\alpha$ . Therefore by 9.611 $Od`u < \omega_{\alpha+1}$ , in other words, $u\,\varepsilon\,F``\omega_{\alpha+1}$, q.e.d.

In order to prove 12.3, we prove at first the following auxiliary theorem:

12.4 From the hypothesis of 12.3 (leaving out closure with respect to C) it follows, that 1.) G is an isomorphism for the triadic relations J_i (i=0,...,8) i.e. (if $G`\alpha$ is abbreviated by $\alpha'$): $J_i`\langle\alpha'\beta'\rangle = [J_i`\langle\alpha\beta\rangle]'$ for $\alpha,\beta\,\varepsilon\,m$, i<9 and 2.) o is closed with respect to the triadic relations J_i.

In outline, the proof runs as follows: By definition of J and the closure property of m, J establishes an isomorphism with respect to S and E between the class of triples $\langle i\alpha\beta\rangle$, i<9, $\alpha,\beta\,\varepsilon\,m$, and m. By G this isomorphism is carried over to an isomorphism between the set t of triples $\langle i\alpha\beta\rangle$, $i<9$, $\alpha,\beta\varepsilon o$, and o. But J likewise defines an isomorphic correspondence between t and some ordinal γ, also with respect to S and E; from this it is inferred by 7.62 that $\gamma=o$ and that J confined to t coincides with the image by G of J confined to $9\times m^2$. But this is what the assertion of the theorem says. The detailed proof is as follows:

Set $j=J\upharpoonright(9\times m^2)$. Then we have $\mathfrak{D}(j)=9\times m^2.\mathfrak{W}(j)\subseteq m$, since m is closed with respect to all the J_i. But also: $m\subseteq\mathfrak{W}(j)$, for suppose $\gamma\varepsilon m$, then $\gamma=J^{\text{‘}}\langle i\alpha\beta\rangle$ for some i,α,β where $\alpha,\beta\varepsilon m$, since m is closed with respect to K_1 and K_2, hence $\gamma\varepsilon\mathfrak{W}(j)$. Therefore $\mathfrak{W}(j)=m$. Moreover,

$$i,k<9\,.\,\alpha,\beta,\gamma,\delta\varepsilon m\,.\,\langle i\alpha\beta\rangle\, S\langle k\gamma\delta\rangle :\supset .\, j^{\text{‘}}\langle i\alpha\beta\rangle<j^{\text{‘}}\langle k\gamma\delta\rangle ,$$

since J has this property, and since for this domain J coincides with j. Therefore $j\,\mathfrak{Isom}_{SE}(9\times m^2,m)$. Now denote by $\bar{j}$ the function into which j is carried over by G, that is, $\bar{j}$ is defined by: $\bar{j}\,\mathfrak{Fn}(9\times o^2).\bar{j}^{\text{‘}}\langle i\alpha'\beta'\rangle=[j^{\text{‘}}\langle i\alpha\beta\rangle]'$, for $\alpha,\beta\varepsilon m$ and $i<9$. This may be written $\bar{j}^{\text{‘}}\langle i\alpha\beta\rangle=[j^{\text{‘}}\langle i\alpha_1\beta_1\rangle]'$, for $\alpha,\beta\varepsilon o$, $i<9$, where $\breve{G}^{\text{‘}}\alpha$ is denoted by α_1. We want to show that $\bar{j}\,\mathfrak{Isom}_{SE}(9\times o^2,o)$. Now: $\mathfrak{D}(\bar{j})=9\times o^2$ and $\mathfrak{W}(\bar{j})=o$, because j has the corresponding properties. Since G is an isomorphism with respect to E it follows by definition 7.8 that $\langle\alpha\beta\rangle\,Le\langle\gamma\delta\rangle.\equiv.\langle\alpha'\beta'\rangle\,Le\langle\gamma'\delta'\rangle$ for $\alpha,\beta,\gamma,\delta\varepsilon m$. Likewise, by definition 7.81, $\langle\alpha\beta\rangle\;R\;\langle\gamma\delta\rangle.\equiv.\langle\alpha'\beta'\rangle\,R\,\langle\gamma'\delta'\rangle$ for $\alpha,\beta,\gamma,\delta\varepsilon m$. It follows then by definition 9.2 that

$$\langle i\alpha_1\beta_1\rangle\,S\langle k\gamma_1\delta_1\rangle\,.\equiv.\,\langle i\alpha\beta\rangle\,S\langle k\gamma\delta\rangle$$

for $\alpha,\beta,\gamma,\delta\varepsilon o$ and $i,k<9$. Now suppose $\alpha,\beta,\gamma,\delta\varepsilon o$; $i,k<9$ and $\langle i\alpha\beta\rangle\,S\langle k\gamma\delta\rangle$. We have then $\langle i\alpha_1\beta_1\rangle\,S\langle k\gamma_1\delta_1\rangle$, which implies, since $j\,\mathfrak{Isom}_{SE}(9\times m^2,m)$, that $j^{\text{‘}}\langle i\alpha_1\beta_1\rangle\,E j^{\text{‘}}\langle k\gamma_1\delta_1\rangle$. Now, since G is an isomorphism with respect to E, we conclude that $[j^{\text{‘}}\langle i\alpha_1\beta_1\rangle]'E[j^{\text{‘}}\langle k\gamma_1\delta_1\rangle]'$, that is, $\bar{j}^{\text{‘}}\langle i\alpha\beta\rangle\,E\bar{j}^{\text{‘}}\langle k\gamma\delta\rangle$. Therefore $\bar{j}\,\mathfrak{Isom}_{SE}(9\times o^2,o)$.

Now define $j_o=J\upharpoonright(9\times o^2)$. Then $\mathfrak{D}(j_o)=9\times o^2$ and $\mathfrak{W}(j_o)$ is some ordinal number γ, since $9\times o^2$ is an S-section of $9\times On^2$. Therefore under J the image must be an E-section of On i.e. an ordinal by 7.21. Hence both $j_o\,\mathfrak{Isom}_{SE}(9\times o^2,\gamma)$ and $\bar{j}\,\mathfrak{Isom}_{SE}(9\times o^2,o)$, but there can exist but one isomorphism of this kind of a set on an ordinal number, by 7.62, hence $\gamma=o$ and $j_o=\bar{j}$. Therefore, $j_o^{\text{‘}}\langle i\alpha'\beta'\rangle=\bar{j}^{\text{‘}}\langle i\alpha'\beta'\rangle=[j^{\text{‘}}\langle i\alpha\beta\rangle]'$, for $\alpha,\beta\varepsilon m$, $i<9$, which is equivalent, by the construction of j_o and j to the statement: $J^{\text{‘}}\langle i\alpha'\beta'\rangle=[J^{\text{‘}}\langle i\alpha\beta\rangle]'$, for $\alpha,\beta\varepsilon m$, $i<9$, which, in turn, is the same as: $J_i^{\text{‘}}\langle\alpha'\beta'\rangle=[J_i^{\text{‘}}\langle\alpha\beta\rangle]'$, for $i=0,\ldots,8$, $\alpha,\beta\varepsilon m$, which is what we set out to prove. That o is closed with respect to the J_i follows immediately from the last equality.

12.4 can be stated symmetrically as follows:

12.5 If $m\subseteq On$, $m'\subseteq On$, m, m' both closed with respect to K_1, K_2 and the J_i as triadic relations and if

$G\,\mathfrak{Isom}_{EE}(m,m')$ then G is an isomorphism for the triadic relations J_i.

The proof is obtained by mapping m and m' on the same ordinal o by 7.7 and then applying 12.4

12.51 The hypothesis of 12.5 implies furthermore:
$\alpha\,\varepsilon\,\mathfrak{W}(J_i).\supset.G'\alpha\,\varepsilon\,\mathfrak{W}(J_i)$ for $\alpha\,\varepsilon\,m$, $i=0,\ldots,8$.

Proof: $\alpha\,\varepsilon\,\mathfrak{W}(J_i)$ implies $\alpha=J_i{}'\langle\beta\gamma\rangle$, $\beta,\gamma\,\varepsilon\,m$ since m is closed with respect to K_1, K_2. Hence $\alpha'=J_i{}'\langle\alpha'\beta'\rangle$ by 12.5; hence $\alpha'\,\varepsilon\,\mathfrak{W}(J_i)$.

Next it will be shown that:

12.6 If m , m' , G satisfy the hypothesis of 12.5 and in addition m and m' are also closed with respect to C then G is an isomorphism for the relations $\hat\alpha\hat\beta(F'\alpha\,\varepsilon\,F'\beta)$ and $\hat\alpha\hat\beta(F'\alpha=F'\beta)$. In other words,
(a) $\alpha,\beta\,\varepsilon\,m.\supset:F'\alpha\,\varepsilon\,F'\beta.\equiv.F'\alpha'\,\varepsilon\,F'\beta'.F'\alpha=F'\beta.\equiv.F'\alpha'=F'\beta'$,
where again $G'\alpha$ is abbreviated by α'.

The scheme of the proof will be to carry out an induction on $\eta=\mathfrak{Max}\{\alpha\beta\}$. We will assume as the hypothesis of the induction that (a) is true for $\alpha,\beta\,\varepsilon\,m$ and $\alpha,\beta<\eta$, and prove it for $\alpha,\beta\,\varepsilon\,m,\mathfrak{Max}\{\alpha\beta\}=\eta$. [Hence the property which is shown by induction to belong to all ordinals η is given by the propositional function:

$$(\alpha,\beta)[\alpha,\beta\,\varepsilon\,m.\eta=\mathfrak{Max}\{\alpha\beta\}:\supset$$
$$:(F'\alpha\,\varepsilon\,F'\beta.\equiv.F'G'\alpha\,\varepsilon\,F'G'\alpha).(F'\alpha=F'\beta.\equiv.F'G'\alpha=F'G'\beta)]\ .$$

This expression is normal; therefore we can apply induction by 7.161.] If $\mathfrak{Max}\{\alpha\beta\}=\eta$ there are 3 possible cases, namely
1.) $\alpha=\beta=\eta$. In this case the equivalences (a) both hold, since, in the first, both members are false, and in the second both are true. The remaining two cases are $\alpha=\eta$, $\beta<\eta$ and $\alpha<\eta$, $\beta=\eta$. Hence what has to be proved is:

$$\left.\begin{array}{l}F'\alpha\,\varepsilon\,F'\eta.\equiv.F'\alpha'\,\varepsilon\,F'\eta'\ ,\\ F'\eta\,\varepsilon\,F'\beta.\equiv.F'\eta'\,\varepsilon\,F'\beta'\ ,\\ F'\eta=F'\beta.\equiv.F'\eta'=F'\beta'\ ,\end{array}\right\}\ \text{for}\ \alpha,\beta<\eta,\ \alpha,\beta\,\varepsilon\,m\ ,$$

under the hypothesis that $\eta\,\varepsilon\,m$ and

$$\begin{array}{ll}\text{I.} & F'\alpha\,\varepsilon\,F'\beta.\equiv.F'\alpha'\,\varepsilon\,F'\beta'\ ,\\ \text{II.} & F'\alpha=F'\beta.\equiv.F'\alpha'=F'\beta'\ ,\end{array}\quad \text{for}\ \alpha,\beta\,\varepsilon\,m\cdot\eta\ .$$

Everything which follows from now on up to the end of the proof of theorem 12.6 [in particular the theorems (1)-(9) on p.57-9] depends on this inductive hypothesis in addition to the hypothesis of theorem 12.6.

The following abbreviations will be convenient: $r=F''m$,

$r_\eta = F``(m \cdot \eta)$, $r' = F``m'$, and $r'_\eta = F``(m' \cdot \eta')$. Hence $r_\eta \subseteq r$ and $r'_\eta \subseteq r'$. Now we can define a one to one mapping H of r_η on r'_η by $H = F|G|F^{-1}$. Because of the inductive hypothesis II, H is one-to-one and $H`x = F`\alpha'$ if $x = F`\alpha$, $\alpha \varepsilon m \cdot \eta$. Because of inductive hypothesis I, H is an isomorphism with respect to E. Note that the hypotheses of theorem 12.6 and the inductive hypothesis are perfectly symmetric in m, m' and η, η' so that whatever is proved from them will also hold if m, η, r, r_η, G, H are interchanged respectively with m', η', r', r'_η, $\breve{G}$, $\breve{H}$.

The next step will be to show that H is an isomorphism for the triadic relation $\hat{z}\hat{x}\hat{y}[z = \langle xy \rangle]$ and the tetradic relation $\hat{z}\hat{u}\hat{v}\hat{w}[z = \langle uvw \rangle]$, and for the Q_i. In order to establish this some preliminary results are needed.

(1) r is closed with respect to the fundamental operations.

Proof: Take $x, y \varepsilon r$, then $x = F`\alpha$, $y = F`\beta$, $\alpha, \beta \varepsilon m$, so that $\mathfrak{F}_i(xy) \varepsilon r$, by 9.31-9.34, since m is closed with respect to the J_i. Therefore $x - y$, $\{xy\}$, $\langle xy \rangle$, $\langle xyz \rangle$, and $x \cdot Q_i``y$ are in $r$ if $x, y, z \varepsilon r$. In particular, it follows that $x \cdot Q_i``\{y\} \varepsilon r$, if $x, y \varepsilon r$.

(2) $x \varepsilon r . \supset . Od`x \varepsilon m$

Proof: $\{x\} \varepsilon r$ by (1), hence there is an $\alpha \varepsilon m$ such that $\{x\} = F`\alpha$. Set $\beta = C`\alpha$. Then $\beta \varepsilon m$, since m is closed with respect to C and $\beta = Od`x$, by definition of C (11.81).

(3) $(x \varepsilon r) . (x \neq 0) : \supset . x \cdot r \neq 0$

Proof: There is an $\alpha \varepsilon m$ such that $x = F`\alpha$. $F`C`\alpha \varepsilon x$, by definition 11.81, but $F`C`\alpha \varepsilon r$, since m is closed with respect to C, hence $x \cdot r \neq 0$.

(3.1) $\{xy\} \varepsilon r . \supset . x, y \varepsilon r$; $\langle xy \rangle \varepsilon r . \supset . x, y \varepsilon r$; $\langle xyz \rangle \varepsilon r . \supset . x, y, z \varepsilon r$.

Proof: It follows from (3) that $\{x\} \varepsilon r . \supset . x \varepsilon r$, because x is the only element of $\{x\}$, also $\{xy\} \varepsilon r . \supset . x, y \varepsilon r$, for, either x or $y \varepsilon r$ by (3); suppose $x \varepsilon r$, then $\{x\} \varepsilon r$ by (1), hence $\{xy\} - \{x\} \varepsilon r$ by (1), so that $\{y\} \varepsilon r$ if $x \neq y$, hence $y \varepsilon r$. By iteration, $\langle xy \rangle \varepsilon r . \supset . x, y \varepsilon r$ and $\langle xyz \rangle \varepsilon r . \supset . x, y, z \varepsilon r$. It follows then that

(4) $y \varepsilon r . \langle yx \rangle \varepsilon Q_i : \supset . x \varepsilon r$ for $i \neq 5$.

Proof: Consider Q_6, the permutation of the ordered pair: If $\langle yx \rangle \varepsilon Q_6$ then $y = \langle uv \rangle$, $\langle vu \rangle = x$ for some u, v. $\langle uv \rangle \varepsilon r$ by assumption, hence $u, v \varepsilon r$ by (3.1) so that $\langle vu \rangle \varepsilon r$, by (1), that is, $x\varepsilon r$. Similarly for the other permutations i.e. Q_7, Q_8. Now consider $Q_4=P_2^{-1}$: assume $y\varepsilon r$, $\langle yx\rangle\varepsilon P_2^{-1}$, then $\langle xy\rangle\varepsilon P_2$ i.e. y is an ordered pair and x its second member, hence $x\varepsilon r$ by (3.1).

There is a weak completeness theorem for r_η :

(5) $x\varepsilon r_\eta . y\varepsilon x :\supset: y\varepsilon r . \supset . y\varepsilon r_\eta$.

Proof: Set $\alpha = Od^{\prime}y$. $\alpha\varepsilon m$, by (2), $Od\ y < Od^{\prime}x < \eta$, by 9.52 hence $\alpha\varepsilon m{\cdot}\eta$, that is, $y\varepsilon r_\eta$.

(6) $y\varepsilon F^{\prime}\eta . y\varepsilon r :\supset . y\varepsilon r_\eta$

Proof: $Od^{\prime}y<\eta$ by 9.52. $Od^{\prime}y\varepsilon m$ by (2) hence $Od^{\prime}y\varepsilon m{\cdot}\eta$ i.e. $y\varepsilon r_\eta$.

(7) $\{xy\}\varepsilon r_\eta . \supset . x,y\varepsilon r_\eta$ and $\langle xy\rangle\varepsilon r_\eta . \supset . x,y\varepsilon r_\eta$;
$\langle xyz\rangle\varepsilon r_\eta . \supset . x,y,z\varepsilon r_\eta$.

Proof: $\{xy\}\varepsilon r$, therefore $x,y\varepsilon r$ by (3.1), hence the result follows, by (5). By iteration it follows that
$\langle xy\rangle\varepsilon r_\eta . \supset . x,y\varepsilon r_\eta$, and similarly for triples.

(8) H is an isomorphism with respect to $\hat{z}\hat{x}\hat{y}[z=\{xy\}]$, $\hat{z}\hat{x}\hat{y}[z=\langle xy\rangle]$, $\hat{z}\hat{x}\hat{y}\hat{t}[z=\langle xyt\rangle]$, and the Q_i $(i=4,5,\ldots,8)$.

[In the sequel $H^{\prime}x$ is abbreviated by $x^{\prime}$. So the prime is an abbreviation for G or H occording as to whether it occurs with a Greek or a Latin letter.]

Proof: Consider $\{xy\}$, we wish to show that
$x,y,z\varepsilon r_\eta . \supset : z=\{xy\} . \equiv . z^{\prime}=\{x^{\prime}y^{\prime}\}$.
Recalling the symmetry of the hypotheses, and that $x,y,z\varepsilon r_\eta$ is equivalent to $x^{\prime},y^{\prime},z^{\prime}\varepsilon r_\eta^{\prime}$, it is obvious that it is sufficient to prove implication in one direction, in order to establish the equivalence. We prove implication from right to left; $z^{\prime}=\{x^{\prime}y^{\prime}\}$ implies $x^{\prime}\varepsilon z^{\prime}$ and $y^{\prime}\varepsilon z^{\prime}$, hence, since H is an isomorphism with respect to E, $x\varepsilon z$, and $y\varepsilon z$, i.e., $\{xy\}\subseteq z$. We have then only to show that $z-\{xy\}=0$. Since $x,y,z\varepsilon r$, $z-\{xy\}\varepsilon r$ by (1), hence by (3), if $z-\{xy\}\neq 0$, there is a $u\varepsilon r$ such that $u\varepsilon[z-\{xy\}] . u\varepsilon z$, and $z\varepsilon r_\eta$, hence, by (5), $u\varepsilon r_\eta$. $u\varepsilon z$, $u\neq x$, and $u\neq y$, hence $u^{\prime}\varepsilon z^{\prime}$, $u^{\prime}\neq x^{\prime}$, and $u^{\prime}\neq y^{\prime}$, because H is one to one and isomorphic for E. But this means $z^{\prime}\neq\{x^{\prime}y^{\prime}\}$, contrary to assumption.

To establish that H is an isomorphism for $z=\langle xy\rangle$ it must be shown that $x,y,z\varepsilon r_\eta . \supset : z=\langle xy\rangle . \equiv . z^{\prime}=\langle x^{\prime}y^{\prime}\rangle$. Again it is sufficient to establish implication in one direction. Assume $z=\langle xy\rangle$. It follows that $z=\{uv\}$, where $u=\{xx\}$ and $v=\{xy\}$. By (7), $u,v\varepsilon r_\eta$, hence, forming $z^{\prime}$, $u^{\prime}$, $v^{\prime}$, $x^{\prime}$, $y^{\prime}$, it follows that $v'=\{x'y'\}$, $u'=\{x'x'\}$, and $z'=\{u'v'\}$, that is, $z'=\langle x'y'\rangle$.

For the ordered triple, assume $z=\langle xyt\rangle$, then $z=\langle xs\rangle$, where $s=\langle yt\rangle$; $t,s\varepsilon r_\eta$, by (7), since $z\varepsilon r_\eta$, therefore $z'=\langle x's'\rangle$, $s'=\langle y't'\rangle$, that is, $z'=\langle x'y't'\rangle$.

Consider now $Q_5=P_2$; we must show that

$$x,z\varepsilon r_\eta .\supset:\langle xz\rangle\varepsilon P_2.\equiv.\langle x'z'\rangle\varepsilon P_2 .$$

As usual, only the implication in one direction is necessary. Assume $\langle xz\rangle\varepsilon P_2$, then there is a y such that $z=\langle yx\rangle$; by (7) $y\varepsilon r_\eta$, therefore $z'=\langle y'x'\rangle$ by (8), that is, $\langle x'z'\rangle\varepsilon P_2$. Now since H is an isomorphism with respect to P_2, H must be an isomorphism also with respect to $Q_4=P_2^{-1}$.

There remain only the permutations Q_6, Q_7, Q_8. Consider Q_6, for example. Assume $\langle xy\rangle\varepsilon Q_6$, then there exist u and v such that $x=\langle uv\rangle$ and $y=\langle vu\rangle$. Since $x,y\varepsilon r_\eta$, it follows by (7) that $u,v\varepsilon r_\eta$, hence $x'=\langle u'v'\rangle$ and $y'=\langle v'u'\rangle$ by (8), that is, $\langle x'y'\rangle\varepsilon Q_6$. The proofs are similar for Q_7 and Q_8.

Now consider the three relations which must be proved to establish the induction, namely,

$$(9)\quad \left.\begin{array}{l} 1.\ F`\alpha\varepsilon F`\eta.\equiv.F`\alpha'\varepsilon F`\eta' ,\\ 2.\ F`\eta\varepsilon F`\beta.\equiv.F`\eta'\varepsilon F`\beta' ,\\ 3.\ F`\eta=F`\beta.\equiv.F`\eta'=F`\beta' ,\end{array}\right\}\ \text{for}\ \alpha,\beta\varepsilon m\cdot\eta .$$

We shall show now that it is sufficient to prove the first of these three relations. Let us assume then that the first is true, and prove the third. Assume that $F`\eta\neq F`\beta$. Then either $F`\eta-F`\beta\neq 0$ or $F`\beta-F`\eta\neq 0$, and $[F`\eta-F`\beta]\varepsilon r$, $[F`\beta-F`\eta]\varepsilon r$ by (1). Hence by (1) and (3) there is a $u\varepsilon r$ such that either $u\varepsilon[F`\eta-F`\beta]$ or $u\varepsilon[F`\beta-F`\eta]$. Therefore $u\varepsilon F`\eta$ or $u\varepsilon F`\beta$, hence, in both cases $u\varepsilon r_\eta$, by (6) and (5), since $F`\beta\varepsilon r_\eta$ . Let us now assume $u\varepsilon[F`\eta-F`\beta]$ , then $u\varepsilon F`\eta$ and $\sim(u\varepsilon F`\beta)$ . Hence by the inductive hypothesis II, we have $\sim(u'\varepsilon F`\beta')$, but also $u'\varepsilon F`\eta'$ , because we have assumed (9) 1. to be true. Therefore $F`\eta'-F`\beta'\neq 0$ . Suppose, on the other hand, that $u\varepsilon[F`\beta-F`\eta]$ , then $u\varepsilon F`\beta$ and $\sim(u\varepsilon F`\eta)$ . Exactly as above, we have $u'\varepsilon F`\beta'$ and $\sim(u'\varepsilon F`\eta')$ , that is, $F`\eta'\neq F`\beta'$ . Thus we have shown $F`\eta\neq F`\beta.\supset.F`\eta'\neq F`\beta'$ and the inverse follows by symmetry reasons as usual.

We have now established that the third relation of (9) follows from the first. Now we derive the second from the first and third. Assume that $F`\eta\varepsilon F`\beta$; set $\alpha=\text{Od}`F`\eta$. By 9.52 $\alpha<\beta<\eta$ and by (2) $\alpha\varepsilon m\cdot\eta$. $F`\eta=F`\alpha$ and therefore $F`\alpha\varepsilon F`\beta$; from $F`\eta=F`\alpha$ it follows by (9) 3. that $F`\eta'=F`\alpha'$, moreover $F`\alpha'\varepsilon F`\beta'$, by the inductive hypothesis I, hence $F`\eta'\varepsilon F`\beta'$, that is, $F`\eta\varepsilon F`\beta.\supset.F`\eta'\varepsilon F`\beta'$ and the inverse implication by reasons of symmetry. Therefore it is sufficient to show (9) 1. and by symmetry reasons it is sufficient to show:

$F`\alpha\varepsilon F`\eta . \supset . F`\alpha'\varepsilon F`\eta'$ for $\alpha\varepsilon m\cdot\eta$.

So we assume $F`\alpha\varepsilon F`\eta$, and consider separate cases according to the index i such that $\eta\varepsilon\mathfrak{W}(J_i)$.

1. Suppose $\eta\varepsilon\mathfrak{W}(J_o)$; by 12.51, $\eta'\varepsilon\mathfrak{W}(J_o)$, hence $F`\eta=F``\eta$ and $F`\eta'=F``\eta'$, by 9.35, so that both members of the equivalence (9) 1. are true, hence trivially equivalent.

2. Suppose $\eta\varepsilon\mathfrak{W}(J_1)$. Then $\eta=J_1`\langle\beta\gamma\rangle$ , where $\beta,\gamma\varepsilon m$ (by the closure property of m ) and $\beta,\gamma<\eta$ , by 9.25. Also $\eta'=J_1`\langle\beta'\gamma'\rangle$ by 12.5 so that 9.31 gives: $F`\eta=\{F`\beta F`\gamma\}$ and $F`\eta'=\{F`\beta' F`\gamma'\}$. Suppose $F`\alpha\varepsilon F`\eta$, then $F`\alpha=F`\beta$ or $F`\alpha=F`\gamma$, therefore, by the inductive hypothesis I, $F`\alpha'=F`\beta'$, or $F`\alpha'=F`\gamma'$, that is, $F`\alpha'\varepsilon\{F`\beta' F`\gamma'\}$ , in other words, $F`\alpha'\varepsilon F`\eta'$.

3. If $\eta\varepsilon\mathfrak{W}(J_2)$ then we have as before, $\eta=J_2`\langle\beta\gamma\rangle$ and $\eta'=J_2`\langle\beta'\gamma'\rangle$, $\beta,\gamma\varepsilon m\cdot\eta$. By 9.32, $F`\eta=E\cdot F`\beta$ and $F`\eta'=E\cdot F`\beta'$. If $F`\alpha\varepsilon F`\eta$, then $F`\alpha\varepsilon F`\beta$ and $F`\alpha\varepsilon E$ . It follows that $F`\alpha'\varepsilon F`\beta'$ , by the hypothesis I of the induction. From $F`\alpha\varepsilon E$ it follows that $F`\alpha=\langle xy\rangle$ and $x\varepsilon y$ for some x,y ; $F`\alpha\varepsilon r_\eta$, hence $x,y\varepsilon r_\eta$, by (7), therefore $F`\alpha'=\langle x'y'\rangle$ , by (8), and $x'\varepsilon y'$ , that is, $F`\alpha'\varepsilon E$. Hence $F`\alpha'\varepsilon E\cdot F`\beta'$, in other words, $F`\alpha'\varepsilon F`\eta'$.

4. If $\eta\varepsilon\mathfrak{W}(J_3)$, we get in the same fashion, by 9.33, $F`\eta=F`\beta-F`\gamma$ and $F`\eta'=F`\beta'-F`\gamma'$, $\beta,\gamma\varepsilon m\cdot\eta$. Assume $F`\alpha\varepsilon F`\eta$, and the inductive hypothesis I applied to $F`\alpha$, with $F`\beta$ and $F`\gamma$ gives $F`\alpha'\varepsilon F`\eta'$ immediately.

5. Suppose $\eta\varepsilon\mathfrak{W}(J_i)$, i=4,6,7,8 . As above, $\eta=J_i`\langle\beta\gamma\rangle$ , $\eta'=J_i`\langle\beta'\gamma'\rangle$, $\beta,\gamma\varepsilon m\cdot\eta$, so that $F`\eta=F`\beta . Q_i``F`\gamma$ and $F`\eta'=F`\beta'\cdot Q_i``F`\gamma'$ , by 9.34. Now assume $F`\alpha\varepsilon F`\eta$ , that is, $F`\alpha\varepsilon F`\beta$ and $F`\alpha\varepsilon Q_i``F`\gamma$ . It follows that $F`\alpha'\varepsilon F`\beta'$ ; also by definition 4.52 there is an $x\varepsilon F`\gamma$ such that $\langle F`\alpha\, x\rangle\varepsilon Q_i . x\varepsilon r$ by (4) and $x\varepsilon F`\gamma\varepsilon r_\eta$, hence $x\varepsilon r_\eta$, by (5), so that, by (8) $\langle F`\alpha' x'\rangle\varepsilon Q_i$ , in addition $x'\varepsilon F`\gamma'$, hence $F`\alpha'\varepsilon Q_i``F`\gamma'$ , hence $F`\alpha'\varepsilon F`\eta'$.

6. There remains now only the case $\eta\varepsilon\mathfrak{W}(J_5)$. As before, $F`\eta=J_5`\langle\beta\gamma\rangle$ and $F`\eta'=J_5`\langle\beta'\gamma'\rangle$, that is, $F`\eta=F`\beta\cdot P_2``(F`\gamma)$ and $F`\eta'=F`\beta'\cdot P_2``(F`\gamma')$. Note that $x\varepsilon P_2``y$ is equivalent to $y\cdot P_2``\{x\}\neq 0$. Suppose $F`\alpha\varepsilon F`\eta$, then $F`\alpha\varepsilon F`\beta$, and $F`\alpha\varepsilon P_2``F`\gamma$, that is, $F`\gamma\cdot\breve{P}_2``\{F`\alpha\}\neq 0$ . $F`\alpha\varepsilon r$ and $F`\gamma\varepsilon r$ , hence by (1) $[F`\gamma\cdot\breve{P}_2``\{F`\alpha\}]\varepsilon r$ , therefore by (3) there is a $u\varepsilon r$ such that $u\varepsilon F`\gamma . u\varepsilon\breve{P}_2``\{F`\alpha\}$ . Then by (5) $u\varepsilon r_\eta$ ; since $u\varepsilon F`\gamma$ and $\langle uF`\alpha\rangle\varepsilon\breve{P}_2$, it follows that $u'\varepsilon F`\gamma'$ and $\langle u'F`\alpha'\rangle\varepsilon\breve{P}_2$ by (8), that is, $F`\alpha'\varepsilon P_2``(F`\gamma')$ , therefore, since $F`\alpha'\varepsilon F`\beta'$ , it follows that $F`\alpha'\varepsilon F`\eta'$. This concludes the proof of 12.6.

12.3 follows immediately from 12.6, since if m, o satisfy the hypothesis of 12.3, o must be closed with respect to J,[1] by 12.4 and with respect to C, K_1, K_2 (because $K_1{}^{\backprime}\alpha \leq \alpha$, $K_2 \alpha \leq \alpha$ by 9.25 and $C^{\backprime}\alpha \leq \alpha$ by definition). Hence m, o satisfy the hypothesis of 12.6.

But on p. 54 it was shown that 12.2 follows from 12.3. So it is proved that the generalised continuum-hypothesis is a consequence of Σ and the additional axiom V=L ,

Q. E. D.

See Note 10 of Notes added to the second printing, p. 69.

APPENDIX

The following list is a continuation of the list of p.13 and shows by the method explained there that all notions and operations for which special symbols are introduced in these lectures (except only $\simeq$ and $\mathfrak{We}$) are normal.

4.1 $Z \varepsilon X \times Y . \equiv . (\exists uv)[\langle uv\rangle = Z . u \varepsilon X . v \varepsilon Y]$

4.11 $Z \varepsilon X^2 . \equiv . Z \varepsilon X \times X$ (similarly for X^3)

4.2 $\mathfrak{Rel}(X) . \equiv . X \subseteq V^2$ (similarly for $\mathfrak{Rel}_3$)

4.4, 4.41, 4.411 $Z \varepsilon \mathfrak{Cnv}(X) . \equiv . (\exists u,v)[\langle uv\rangle = Z . \langle vu\rangle \varepsilon X]$
(similarly for $\mathfrak{Cnv}_2$, $\mathfrak{Cnv}_3$)

4.42 $Z \varepsilon X + Y . \equiv : Z \varepsilon X . \vee . Z \varepsilon Y$

4.43 $Z \varepsilon X - Y . \equiv : Z \varepsilon X . \sim (Z \varepsilon Y)$

4.45 $Z \varepsilon \mathfrak{W}(X) . \equiv . Z \varepsilon \mathfrak{D}(\mathfrak{Cnv}(X))$

4.5 $Z \varepsilon X \upharpoonright Y . \equiv . Z \varepsilon X \cdot (V \times Y)$ (similarly for $\restriction$ (4.512))

4.52 $Z \varepsilon X``Y . \equiv . Z \varepsilon \mathfrak{W}(X \upharpoonright Y)$

4.53 $Z \varepsilon X | Y . \equiv . (\exists u,v,w)[Z = \langle uw\rangle . \langle uv\rangle \varepsilon X . \langle vw\rangle \varepsilon Y]$

4.6 $\mathfrak{Un}_2(X) . \equiv : \mathfrak{Un}(X) . \mathfrak{Un}(\mathfrak{Cnv}(X))$

4.61 $\mathfrak{Fnc}(X) . \equiv : \mathfrak{Rel}(X) . \mathfrak{Un}(X)$

4.63 $X \mathfrak{Fn} Y . \equiv : \mathfrak{Fnc}(X) . \mathfrak{D}(X) = Y$

4.65 $Z \varepsilon X`Y . \equiv . (\exists u)[Z \varepsilon u . (v)(\langle vY\rangle \varepsilon X . \equiv . v = u)]$

4.8 $Z \varepsilon \mathfrak{S}(X) . \equiv . (\exists u)[Z \varepsilon u . u \varepsilon X]$ (the same proposition holds for $\mathfrak{Max}$ and $\mathfrak{Lim}$)

4.84 $Z \varepsilon \mathfrak{P}(X) . \equiv : \mathfrak{M}(Z) . Z \subseteq X$

6.1 $X \mathfrak{Con} Y . \equiv . Y^2 \subseteq X + \mathfrak{Cnv}(X) + I$

6.3 $X \mathfrak{Sect}_Z Y . \equiv : Y \cdot Z``X \subseteq X . X \subseteq Y$

6.31 $Z \varepsilon \mathfrak{Seg}_T(XY) . \equiv . Z \varepsilon X \cdot T``\{Y\}$

6.4 $Z \, \mathfrak{Isom}_{P,Q}(X,Y) . \equiv :: . \mathfrak{Rel}(Z) . \mathfrak{Un}_2(Z) . \mathfrak{D}(Z) = X . \mathfrak{W}(Z) = Y .$
$(u,v)[u,v \varepsilon X . \supset . (\langle uv\rangle \varepsilon P . \equiv . \langle Z`u Z`v\rangle \varepsilon Q)]$

6.5 $\mathfrak{Comp}(X) . \equiv . \mathfrak{S}(X) \subseteq X$

6.6 $\mathfrak{Ord}(X) . \equiv : \mathfrak{Comp}(X) . E \mathfrak{Con} X$

6.61 $\mathfrak{O}(X) . \equiv : \mathfrak{Ord}(X) . \mathfrak{M}(X)$
$\alpha, \beta, \gamma, \ldots,$ are normal variables since their range is the class On.

6.63, 6.64 $X < Y . \equiv . X \varepsilon Y$; $X \leq Y . \equiv : X \varepsilon Y . \vee . X = Y$

7.4 $Z \varepsilon X \dotplus 1 . \equiv : Z \varepsilon X . \vee . [Z = X . \mathfrak{M}(Z)]$

8.12 $X \dot{\simeq} Y . \equiv : . (\exists u)[\mathfrak{Rel}(u) . \mathfrak{Un}_2(u) . X = \mathfrak{D}(u) . Y = \mathfrak{W}(u)]$

8.2 $Z \varepsilon \overline{\overline{X}} . \equiv : Z \varepsilon On . (\alpha)[\alpha \dot{\simeq} X . \supset . Z \varepsilon \alpha]$

8.48, 8.49 $\mathfrak{Fin}(X) . \equiv . (\exists \alpha)[\alpha \varepsilon \omega . X \dot{\simeq} \alpha]$, $\mathfrak{Inf}(X) . \equiv . \sim \mathfrak{Fin}(X)$

9.1 $Z \varepsilon \mathfrak{F}_1(XY) . \equiv . Z \varepsilon \{XY\}$, $Z \varepsilon \mathfrak{F}_2(XY) . \equiv . Z \varepsilon E \cdot X$,
(similarly for $\mathfrak{F}_3, \ldots, \mathfrak{F}_8$)

9.41 $\mathfrak{L}(X) . \equiv : X \subseteq L . (u)[u \varepsilon L \supset u \cdot X \varepsilon L]$

See Note 8 of Notes added to the second printing, p. 68.

INDEX

I. Special Symbols

II. Letters and Combinations of Letters

[Note that the letters C, F, R, S also occur as variables before their respective definitions as constants. Operations and notions are denoted in general by German letters, classes by Latin letters.]

Variables:

$X,Y,Z,\ldots,A,B,C,\ldots$, for classes
$x,y,z,\ldots,a,b,c,\ldots$, for sets
$\alpha,\beta,\gamma,\ldots$, for ordinal numbers
$i,k,\ldots$, for integers
$\bar{X},\bar{Y},\ldots,\bar{A},\bar{B},\ldots$, for constructible classes
$\bar{X},\bar{Y},\ldots,\bar{a},\bar{b},\ldots$, for constructible sets

III. Technical Terms

Notes Added to the Second Printing

Note 1 (to p. 1). In particular this stronger proposition implies that there exists a projective well-ordering of the real numbers (to be more exact, one whose corresponding set of pairs is a PCA-set in the plane). This follows by considering those pairs of relations s,e between integers which, for some $\gamma < \omega_1$ are isomorphic with the pair of relations $<, \hat{\alpha}\hat{\beta}$ $[F'\alpha \varepsilon F'\beta]$ confined to γ. The class M of these pairs s,e can also be defined directly (i.e. without reference to the previously defined F) by requiring that (1) s is to be a well-ordering relation for the integers, and (2) e, with respect to the well-ordering s, satisfies certain recursive postulates, which are the exact analogues of the postulates by which F is defined (cf. Dfn 9.3). The definition of M, in this form, contains quantifiers only for integers and sets of integers (i.e. real numbers) which ensures the projective character of the object defined and makes it possible to determine its projective order by counting the "changes of sign" of the quantifiers for real numbers occurring. In terms of M a projective well-ordering of the real numbers (of the order mentioned) can then be defined. As to consequences of this state of affairs cf. A. Kuratowski, Fund. Math. 36 (1949).

Note 2 (to p. 6). The term now in use for Ax. C4 is "axiom of replacement."

Note 3 (to p. 6, footnote 4). In this form axiom D, under the name of "Fundierungsaxiom," was first formulated by E. Zermelo in the paper quoted on p. 1 of these lectures.

Note 4 (to p. 6). Using Dfn 4.65 the axiom of choice can be expressed in the following form, equivalent with axiom E: There exist classes A for which: $x \varepsilon y \supset A‘y \varepsilon y$.

Note 5 (to p. 11). One may wish, for aesthetic reasons, that in analogy with axiom A2, one should have $\langle XY \rangle \varepsilon Z \supset . \mathfrak{M}(X) . \mathfrak{M}(Y)$. This can easily be accomplished by replacing in Dfn 3.1 $u = X$ by: $u = X \vee [\mathfrak{Pr}(X) . u \varepsilon X]$, and likewise $u = Y$ by: $u = Y \vee [\mathfrak{Pr}(Y) . u \varepsilon Y]$. If this definition is adopted $\mathfrak{M}(A‘x)$ can be dropped in Dfn 4.65. Otherwise it is indispensable as was noted by Mr. W. L. Duda who called my attention to its omission in the first edition. It is not difficult to define $\{XY\}$ in such manner that 1.13 also holds for proper classes, but since there is never any occasion of making use of this fact there is no point in doing so.

Note 6 (to p. 11, footnote 6). A similar remark applies to many other concepts which by their usual definition are meaningful only for certain classes, e.g. $\mathfrak{Cnv}$, $\mathfrak{Cvn}$, etc. only for classes of pairs; $\mathfrak{Max}$, $\mathfrak{Lim}$ only for sets of ordinals (with or without greatest element resp.) etc. All that is aimed at in the subsequent definitions is that, for those arguments for which, by their usual definitions, the concepts defined are meaningful, the definitions given should agree with the usual ones. For $\mathfrak{Max}$ and $\mathfrak{Lim}$ e.g. this requirement can be satisfied by setting them both equal to $\mathfrak{S}$ (Cf. Dfn 7.31).

Note 7 (to p. 12). The term "concept" only fits to notions and operations. Special classes should rather be called "objects."

Note 8 (to p. 21, 30, 62). The statements made after Dfn 6.2 and 8.1, and on p. 62 to the effect that $\mathfrak{We}$ and $\sim$ are not normal are incorrect, if normality is defined as on p. 12. According to this definition normality of a concept has nothing to do with the way in which it is defined but only depends on its extension. Therefore all that, prima facie, can be said about $\mathfrak{M}\mathfrak{e}$ and $\simeq$ is that they cannot be proved to be normal by the method applied to the other concepts on p. 62. They can however be proved to be normal in a different way, provided the axiom of choice is assumed. For, under this assumption, it can be proved that

$$X \simeq Y . \equiv . X \simeq' Y \vee : \mathfrak{Br}(X) . \mathfrak{Br}(Y)$$

(Cf. J. v. Neumann, J. reine angew. Math. 160). Moreover U can be replaced by u in Dfn 6.2 because the existence of a class without first element implies the existence in it of a descending sequence of type ω. The latter proof requires the singling out of one element in every non empty class, which however can be accomplished by considering, in every class, the subset of elements of lowest "Stufe" (in the sense of J. v. Neumann, l.c., p. 238).

Note 9 (to p. 30). Dfn 8.2, for the case that $\mathfrak{Br}(X)$, is justified by J. v. Neumann's result (concerning $\simeq$) quoted in Note 8.

Note 10 (to p. 61). The above given consistency proof can easily be extended for the case that stronger axioms of infinity are added (e.g. the axiom of the existence of unaccessible numbers, or others given by P. Mahlo, Ber. Verh. Sachs. Ges. Wiss. 63 (1911), 65 (1913)), for the simple reason that all these axioms of infinity imply their own relativized form. A similar remark also applies to extensions of the system Σ by other axioms suggested by the intuitive meaning of the primitive terms. (Added 1966: This holds for the axioms of infinity and other additional axioms known at that time [1951].)

Added August 1965

Note 11. In the past few years decisive progress in the foundations of set theory has been achieved by Paul J. Cohen, who invented a powerful method for constructing denumerable models. This method yields answers to several most important consistency questions. In particular Paul J. Cohen in (2) has proved that Cantor's continuum hypothesis is unprovable from the axioms of set theory (including Mahlo or Levy type axioms of infinity), provided these axioms are consistent. The value that can consistently be assigned to $2^{\aleph_\alpha}$ turns out to be almost completely arbitrary. See (2), (9) and (3).

Note 12. Other quite important progress has been made in the area of axioms of infinity, namely:

1. Mahlo's axioms of infinity have been derived from a general principle regarding the totality of sets, that was first introduced by A. Levy in (7). It gives rise to a hierarchy of different precise formulations. One, given by P. Bernays, implies all of Mahlo's axioms [see (1)].

2. Propositions which, if true, are extremely strong axioms of infinity of an entirely new kind have been formulated and investigated as to their consequences and mutual implications in (10) and (11) and the papers cited there. In contradistinction to Mahlo's axioms the truth (or consistency) of these axioms does not immediately follow from "the basic intuitions underlying abstract set theory" [see (10), p. 134], nor can it, as of now, be derived from them. However, the new axioms are supported by rather strong arguments from analogy, such as the fact that they are implied by the existence of generalizations of Stone's representation theorem to Boolean algebras with operations on infinitely many elements. As was conjectured in a general way in (5), p. 520, one of the new axioms implies the existence of non-constructible sets [see (8)]. Whether one of them implies the negation of the generalized continuum hypothesis has not yet been determined.

NOTES

Note 13. A general discussion of Cantor's continuum problem and its relationship to the foundations of set theory is given in (5) and (6).

Note 14. A slightly different version of the consistency proof given in these lectures, which exhibits more clearly the basic idea of it, is outlined in (4).

BIBLIOGRAPHY

(1) P. Bernays, Essays on the Foundations of Mathematics dedicated to Prof. A. H. Fraenkel, ed. by Y. Bar-Hillel, Jerusalem 1961, p. 3.

(2) Paul J. Cohen, Proc. Natl. Ac. Sci. 50 (1963), p. 1143 and 51 (1964), p. 105. See also the notes of his lectures at Harvard University in Spring 1965 (to be published in the near future).

(3) William B. Easton, Doctoral dissertation, Princeton, 1964, Notices Am. Math. Soc., 11 (1964), p. 205.

(4) K. Gödel, Proc. Natl. Ac. Sci. 25 (1939), p. 220. For the correction of some misprints and confusions of letters that occurred in this paper see (5) footnote 23 or (6) foot note 24.

(5) _____, Am. Math. Monthly 54 (1947), p. 515. Note that in the self references to footnotes that occur in the footnotes 2, 20 and 23 of this paper the numbers 13, 17, 19, 21 should be replaced by 15, 20, 23, 26 respectively.

(6) _____, Philosophy of Mathematics, ed. by P. Benacerraf and H. Putnam, 1964, p. 258. This is a revised and enlarged version of (5).

(7) Azriel Levy, Pac. J. of Math. 10 (1960), p. 233.

(8) Dana Scott, Bull. Pol. Ac. Sci., Ser. sci. math. astr. phys. 9 (1961), p. 521.

(9) Robert M. Solovay, Notices Am. Math. Soc. 10 (1963), p. 595.

(10) A. Tarski, Logic, Methodology and Philosophy of Science (Proceedings of the 1960 International Congress) ed. by E. Nagel, P. Suppes and A. Tarski, Stanford Univ. Press, 1962, p. 125.

(11) H. J. Keisler and A. Tarski, Fund. Math. 53 (1964), p. 225-308.

Milton Keynes UK
Ingram Content Group UK Ltd.
UKHW010247081023
430060UK00001B/39

9 780691 079271